MAGNOLIAS, SWEET TEA, AND EXHAUST

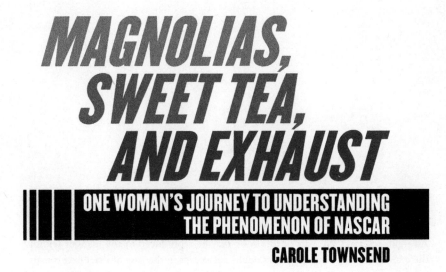

MAGNOLIAS, SWEET TEA, AND EXHAUST

ONE WOMAN'S JOURNEY TO UNDERSTANDING THE PHENOMENON OF NASCAR

CAROLE TOWNSEND

SPORTS PUBLISHING

Copyright © 2014 by Carole Townsend

All rights reserved. No part of this book may be reproduced in any manner without
the express written consent of the publisher, except in the case of brief excerpts in
critical reviews or articles. All inquiries should be addressed to Sports Publishing, 307
West 36th Street, 11th Floor, New York, NY 10018.

Sports Publishing books may be purchased in bulk at special discounts for sales pro-
motion, corporate gifts, fund-raising, or educational purposes. Special editions can also
be created to specifications. For details, contact the Special Sales Department, Sports
Publishing, 307 West 36th Street, 11th Floor, New York, NY 10018 or sportspub-
books@skyhorsepublishing.com.

Sports Publishing® is a registered trademark of Skyhorse Publishing, Inc.®, a Delaware
corporation.

Visit our website at www.sportspubbooks.com.

10 9 8 7 6 5 4 3 2 1

Library of Congress Cataloging-in-Publication Data is available on file.

ISBN: 978-1-61321-691-0

Printed in the United States of America

All interior photos by Steve Bowers

Contents

Acknowledgments

Thank you to the following people and organizations for their gracious assistance and contributions to *Magnolias, Sweet Tea, and Exhaust.*

Mr. Ed Clark and the staff at Atlanta Motor Speedway, for their time, conversation, and the amazing accommodations during "The Country's Biggest Labor Day Party."

My dear friends Frances and Steve Bowers, who tutored me from the beginning in the ways of NASCAR and who shared several race experiences with me during this journey. Frances, you're amazing.

All of the drivers (from those just starting out to the Hall of Famers), team owners, mechanics, and crew chiefs who took the time to talk with me and teach me.

Mr. Doug Allen, for often guiding me, having me on his show, and laughing at me when I started to take myself too seriously.

All of the many fans who shared their passion for NASCAR with me.

Driver Jay Foley, who, at 165 mph, schooled me in the adrenaline rush of speed and kept me alive at the same time.

The gracious Vaughan family, for sharing their peaceful beach home with me for the writing of much of this book.

And of course my husband, Marc, for being my biggest supporter, my most genuine encouragement, and a great sport.

Introduction

An airplane was about to land on our house. In fact from the sound of it, I believed it would make contact any minute now. I should have been alarmed, I know, but I couldn't seem to move. *Get up! Get out of the way! Move!* My mind was doing its best to shake me awake, to snap me out of my foggy sleep, to find safety at once. The roaring was insistent—loud, louder, then deafening, then zooming off into the distance, not quite as urgent, now faraway, but it was coming back. I could hear it. I opened my eyes just a bit, and nothing around me looked familiar. My panic notched up a bit higher. I closed my eyes, then tried opening them again, all the while the powerful roar growing stronger, more insistent. I braced myself for impact. Then one thought hurried to the forefront of my racing thoughts, overpowering all the others. *Where am I?* The thought sprinted through my head in flailing terror, the question furiously searching for an answer as the fog surrounding my brain began to lift and I started to get my bearings. I was not in my bed, in my room, in my house. *I am on the bottom bunk in a tiny room, there is someone sleeping beside me, my head is killing me, and that plane is coming fast.*

Gingerly fanning my fingers out, I found the edge of the bed and pulled myself up to a sitting position. My head cleared a bit more; the alarm subsided, and the felt-covered pounding in the back of my brain elbowed past the panic to take a front row seat right behind my eyes. I looked over at the man sleeping beside me, snoring ever so slightly. Yes, that was my husband (of course it was), and he was apparently oblivious to the unfamiliar and insistent noises coming from the other side of the thin wall that defined our sleeping quarters. *Cars. That is not the sound of an airplane. What I'm hearing is the sound of race cars.* I sat still, closing my eyes as I sorted out my situation and the events that had gotten me there. *We are in a camper. OK, yes, I remember that, and we are in Hampton, Georgia, at Atlanta Motor Speedway.* Now we were getting somewhere.

We had a great time last night, which explained my dull but unrelenting headache that refused to be ignored. It pounded on the backs of my eyes like a child pitching a whopper of a temper tantrum.

Ah, yes. Fully awake, I reached over my sleeping husband and used one finger to pull back the venetian blinds covering the window over his head. Sunlight darted in and took a quick stab at my eyes before I squinted, closing them tight in protest. The initial jolt of pain diminishing, I pulled the blinds back again and looked outside. I was able to label the sounds that had crept into my deep sleep and shaken me awake, from a distance at first, but now right outside my window. There was the sound of those cars, yes, but there were other noises too. Rumbling, shouting, and . . . wait, was that the sound of children screaming and laughing? It was.

The rumbling was the sound of a few straggling eighteen-wheelers idling and inching forward reluctantly, feeling their way along the short road and grumbling at the effort that it took. The ground shook and groaned with their weighty movement. I got up on my knees and looked as far to the left and right as I could through the small window. I felt like we had been plopped right down in the middle of Oz during the night; when we had gone to bed, there was some activity going on—mostly, more performers and their entourages settling in and setting up to do what they do, but a few campers had been parked here in the middle of the arena with us too. This morning, the entire infield of Atlanta Motor Speedway had been transformed into a gigantic, industrious beehive. Million-dollar RVs and old dilapidated school buses—and everything in between—crept slowly past us, looking for the spot that would be their home for the weekend. I eased myself out of bed, hastily threw on my bathrobe, and tiptoed through the kitchen so as not to wake the other couple asleep on the far end of the camper. I opened the door to the outside ever so carefully and stepped out onto the top step, gawking at the activity that surrounded our neat and tidy little camper. To my right, a car whizzed by at about 120 mph, give or take a hundred miles, and the sound was both deafening and exciting. To my left, the children of some of the drivers and mechanics played happily, their mothers, grandparents, and caretakers busily setting up their camping areas with chairs, picnic tables, and other such camping paraphernalia.

Just beyond our camping area and surrounded by a high wrought-iron fence was where the real activity was winding up, soon to be at a fevered pace with only a few fortunate fans and other onlookers allowed inside for an up-close look at what goes on in those garages. Behind that fence was the nerve center of the weekend's main events. In just about an hour, I'd be right there in the middle of all the action, flashing my track credentials at the security guards and feeling that secret sense of belonging that having such authorization gives the flasher. My panic was replaced with excitement, and I could feel a wide smile spread across my face in spite of my epic, relentless headache.

Hydraulic lifts and air-powered tools huffed and wheezed, and an occasional shout came from a mechanic who, from this angle anyway, looked as though he was being devoured by a slick, shiny race car. Every now and then, he would look out from under the hood of the car and shout something to another mechanic who was working just as intently on the opposite end of the same car. Some of the Sunday drivers (and believe me, in NASCAR, that term means something entirely different than its customary meaning) had not arrived at that point, but their engineers and mechanics were already doing their thing, working their mechanical magic. Far to the left of the garages and just outside the track was the media lot, a parking area packed shoulder-to-shoulder with trucks, vans, towers, and satellite dishes. Wires and cables snaked and snarled in between the metal trees in that electronic jungle, waiting to come alive with data and information. On the other side of the mechanics, the cars, and the hulking trucks was a wall of silver crowned with a glassy row of windows—the steep grandstands of Atlanta Motor Speedway, topped with plush suites.

We had been sleeping at the center of this whole ruckus. "We" happens to be me, my husband (both of us inexperienced in the world of NASCAR), and our dear friends Frances and Steve, who are longtime, dedicated NASCAR fans. We had the privilege of enjoying this vantage point because Mr. Ed Clark, president and general manager of this massive and legendary racetrack, had graciously offered it during an interview I conducted with him months ago, not long after I began researching the puzzling phenomenon of NASCAR for the purposes of writing this book. When I met with him early on in the 2013 season, I mentioned the fact that every fan I had interviewed up to that point had talked about infield camping at Atlanta Motor Speedway with such reverence that I had to wonder, "What's the big deal?" To my delight, Clark asked me if I'd like to experience it for myself, and I jumped at the chance. What better way to fully experience a race than to camp right in the

ATL campers. Note the more upscale RVs at top of hill.

middle of the track and mingle with devout, hardcore fans? How better to understand the sweat and technology and sheer genius of a NASCAR race car than to see the cars and walk through the garages, talk to the drivers and mechanics? Oh yes, this weekend would prove to be a priceless experience. Those coveted camping spots sell out years in advance; there was no way I was going to turn down such an opportunity.

I was standing on the top step of our camper entrance (which, if I may say, was decorated quite festively with Christmas lights and sunny summer flowers), barefoot and in my bathrobe. My hair was wild, not even brushed (which would have only made it worse, anyway), I wore no makeup, and I was standing like this with my mouth agape, trying to take it all in, in the midst of probably ten thousand people (the crowd would eventually swell to about ten times that number of people as the weekend progressed).

A year ago, I would have never set foot outside our bedroom without my makeup on and my hair done—not on a bet, not on a dare, and not even for cold, hard cash. But a year ago, I was not the same woman that I am today.

My head had cleared completely now, though it throbbed in the 90-degree heat and with the cacophony that surrounded us on all sides. Still, my mind had recuperated from my deep sleep, and it had all its facts lined up in a straight, orderly manner. Today was Saturday. Tonight, at 7:30 p.m., some rising stars and some well-known drivers would compete in the NASCAR Nationwide series race right here on this track surrounding our camper. Those nice people we met yesterday were from Kentucky, I remembered that, and they really knew how to throw a party. I remembered most of that.

Someone opened the door behind me, and the inviting smell of coffee, that precious black elixir, the Dark Mother that promised to get me fully awake and make it all better, wafted out on a chilled wave of conditioned air. My friend Frances, dressed for the day and grinning—literally—from ear to ear, said excitedly, "Hurry up! Let's eat! Let's go!"

Frances had already donned her walking shoes, baseball cap, and sunglasses, and slipped the lanyard holding her credentials around her neck, ready to rub elbows with her NASCAR idols. Before both the Saturday and Sunday races this weekend, all four of us would be escorted past security guards, through checkpoints, and into the garages that stowed the impressive cars of drivers like Jeff Gordon, Sam Hornish, Dale Earnhardt Jr., Clint Bowyer, Danica Patrick, the Busch brothers, and more. Frances, a longtime fan of NASCAR and a walking encyclopedia of driver and race information, is as skilled an autograph hunter as I believe I've ever met, and she was primed and ready to go. Breakfast would be a mere formality today; there would be no dallying, I could see that.

This was it, the culmination of nearly a year's worth of work, travel, and research. This weekend would mark the end of one of the most extreme, most unusual, most entertaining journeys I have ever taken in my half-century of life.

I snapped out of my reverie, darted through the kitchen, and unceremoniously jumped onto the bottom bunk in our bedroom, where my husband still snoozed peacefully.

"Honey, wake up!" I said as I excitedly shook him out of his deep sleep. We had some preparing to do; I had waited nearly an entire year for this very event this very weekend, and I intended to soak up every last detail. They didn't call this weekend at Atlanta Motor Speedway the Biggest Labor Day party in the country for nothing, and I wasn't going to miss a thing.

Chapter One

NASCAR—What's the Big Deal?

"Stock car racing has got distinct possibilities for Sunday shows, and we do not know how big it can be if it's handled properly."
– Bill France Sr., founder of NASCAR

This prophetic statement is displayed on a wall at the NASCAR (National Association for Stock Car Auto Racing) Hall of Fame in Charlotte, North Carolina. Bill France Sr., the founder of NASCAR, delivered it at the very first organizational meeting for NASCAR on December 12, 1947, at the Streamline Hotel in Daytona Beach, Florida. The rest, as they say, is history.

France was then a gas station owner, race promoter, and himself a stock car driver. At the aforementioned meeting, he was addressing journalists, drivers, sponsors, and race car owners, all people who believed in his ideas for better organized stock car racing, a sport rather than a haphazard sideshow, in which drivers were respected, one in which they could count on getting paid for driving rather than having to chase down unscrupulous promoters for a purse that sometimes consisted of a little bit of money and one that, sometimes, consisted of livestock and merchandise.

Mr. France, as it turns out, was onto something. Out of that 1947 organizational meeting came an agreement by all in attendance to support France's vision of an organized sport in which there were guidelines and rules, one in which race car drivers were sure to get paid for their efforts. The trifold "rules" brochure printed and circulated soon after that meeting listed "Roadster Specifications" on one panel and covered things like brakes, wheels, axles, bumpers, and guidelines for the appearance of competing cars. While today's rules and guidelines manuals are not available to fans, I can say that I have been told by a reliable source that a recent Sprint Cup Series rule book was just over 170 pages long, and that was some very fine print.

The NASCAR organization started out small, holding races on short dirt and sand tracks. Very quickly, though, it began attracting the best drivers and the best sponsors, and with them came the most devoted fans. While NASCAR sprouted in the South and boasted some of the fastest drivers around, the racing craze quickly spread north and west, and soon there were more than one hundred tracks from coast to coast. Today, there are more than seventy-five million NASCAR fans in the United States alone. Nearly half of those fans are women, a statistic that, quite frankly, astounded me when I first read it. I also read somewhere that NASCAR races draw larger crowds than NFL football, MLB baseball, and NBA basketball finals combined, though the ailing economy has noticeably impacted attendance in recent years. In February 2013, NASCAR put an end to its customary practice of providing attendance figures following every race. However, according to NASCAR's own estimates, more than three and a half million fans attended 2012's thirty-six Sprint Cup races. For the first time in many years, empty seats are sprinkled throughout the crowds of one hundred thousand–plus fans. Still, millions of fiercely loyal race devotees spend more than three billion dollars annually on NASCAR-licensed merchandise and sponsor products.

NASCAR doesn't do anything on a small scale; it's all about speed, big money, and big deals. For instance, as recently as July 2013, NASCAR teamed up with actress Alyssa Milano's female-fan-friendly Touch brand of clothing, aiming to design a fashion-forward line of women's clothing for NASCAR fans. I have no doubt that the deal was a lucrative one, and I can attest to the fact that fashion-conscious female NASCAR fans will appreciate the move.

Television viewership of NASCAR is a goldmine for sponsors and networks. Social media giants Facebook and Twitter buzz with NASCAR fan comments and driver tweets; on race Sundays, the back-and-forth chatter between fans, crew members, and even drivers scrolls like a Wall Street ticker. NASCAR is second only to the NFL among professional sports in terms of TV viewing ratings in the United States.

Here's a surprising piece of information: NASCAR races are broadcast in more than 150 countries and in twenty languages, and races have actually been held in foreign countries, including Mexico, Canada, and Australia, to name a few. According to a February 2013 article written by David Caraviello in NASCAR.com, NASCAR wants to have fans, drivers, and races on the world stage, not just in the United States. If a driver has what it takes to compete in what many say is the toughest motorsport in the world, he and his fans will be welcomed with open arms. Mexico has a series of NASCAR races. Promising European drivers compete in Euro-Racecar, the first official NASCAR series in Europe. The Euro-Racecar NASCAR Touring Series kicked off in April 2012 in France. NASCAR in France? Stock car racing right alongside *fromage* and champagne? *Oui*, it's true.

Joe Balash, NASCAR's International Competition Liaison, has a single, simple goal in mind: "The world population right now is seven billion. We want all of 'em" (David Caraviello, NASCAR.com, Feb. 27, 2013). Yes, NASCAR is all about "big." Balash's lofty goal flies in the face of my old impression of NASCAR, the one I admittedly had before this adventure, which limited the appeal of stock car racing to blue-collar and (dare I say it) redneck fans.

Fortune 500 companies sponsor NASCAR more than any other motorsport, even though the cost of doing so is steep, ranging anywhere from the tens of thousands to the millions of dollars. When I set out on my journey to learn all I could about NASCAR in a single season, I had no idea of the scope and scale and of the kind of money that surrounds the sport, from the billionaire owners, to the millionaire drivers, to the seasoned crew chiefs' and engineers' compensations. The numbers are staggering. NASCAR royalty Brian France and Bruton Smith are billionaires, and that's with a *b*.

Without delving too deeply into the nitty-gritty of specifics, here is what I've learned about why dollars flock to NASCAR as they seem to do, with hefty price tags attached to everything from the cars themselves to the expense of sponsoring a driver. Why does it cost so much to sponsor a NASCAR driver? There are two reasons, I've learned. First, a healthy return on the sponsor's investment is pretty much a given. Second, NASCAR is a pricey sport. For these reasons alone, even the most accomplished drivers are constantly looking to acquire and keep sponsors, and the competition is fierce.

The cars are mind-blowingly expensive to build and run, and at any given time, the top motorsports garages are building more than just one car for a particular driver to race. The most recent research I could find clued me in to the cost of the NASCAR Sprint Cup race car itself, though exact numbers are rarely disclosed and many figures are held close to the chest among mechanics, managers, and others.

To own and operate a Generation 6 race car (the current iteration of NASCAR cars, which debuted at Daytona Beach in 2013 and is currently being tested on tracks across the nation) costs roughly anywhere from $150,000 to $175,000 for each bare chassis. That's

Sponsors can be found just about anywhere if drivers just look hard enough. This is driver Morgan Shepherd's car.

without a powertrain or even any sheet metal. Engine costs can run from $45,000 to $90,000+ each for the season, and a bigger racing shop like Hendrick Motorsports or Richard Childress Racing in North Carolina typically supports four NASCAR teams a year; an engine department with dozens of engineers and mechanics costs dearly, with six-figure salaries not uncommon among the most talented of them. Competition is fierce for the coveted positions in top garages, with only the best of the best landing jobs in those squeaky-clean, high-tech wonder-shops.

Secrecy surrounds the technology spawned by the garages that team with NASCAR drivers, as even the slightest advantage can make the difference between first and last place to a team. State-of-the-art garages (most of them clustered around Charlotte and Mooresville, North Carolina) are staffed not with uneducated, profanity-slinging grease monkeys but rather with engineers and mechanics who have degrees from traditional universities and have completed engineering programs at the masters and PhD level. In fact, the University of North Carolina has a program dedicated solely to engineers who want to work for NASCAR, and those who complete the program come very well trained, but do not come cheap.

I was amazed to learn that teams pay around $20,000 per race just for tires, which cost about $400 apiece and apparently wear out faster than cheap flip flops on hot concrete. The Goodyear Racing Eagle tires (NASCAR allows only one tire manufacturer—Goodyear—to provide tires to its teams) are eleven inches wide and have no tread whatsoever, allowing the road-hungry cars a much better grip as they compete for the lead lap after lap. Interestingly enough, race teams lease these tires from Goodyear, since NASCAR decided that the bigger, better-funded teams had a purchasing advantage over smaller teams. Used tires are returned to Goodyear and recycled.

The chassis of a race car is built for safety and aerodynamics, but certainly not for comfort. The chassis is also expensive, and that's with one seat and no frills like carpet, upholstery, power windows, a lighted passenger-side mirror, or even a gas gauge. It's money well spent, as many of today's drivers have walked away from wrecks that would have killed them years ago. The NASCAR Hall of Fame currently has an entire display, appropriately named *Wrecks! Dramatic Crashes of NASCAR*, which features film, artifacts, and spectacularly preserved wreckage of several race cars.

In the case of each twisted and burned wreck on display, the driver lived to race again another day. That know-how costs money.

Costs to build NASCAR stock cars have risen, according to several mechanics, with the appearance of the fifth generation, NASCAR's "Car of Tomorrow," and the current Gen 6 cars. One small change to a car can have the ripple effect of thousands to possibly millions of dollars for each racing team. NASCAR officials are aware of this financial impact, and for that reason they try to phase in some changes over a period of years to minimize the hardship on teams.

According to a *New York Times* article written by Leo Levine and dated February 22, 2013, a Sprint Cup team can spend as much as

One of the wrecks displayed at Hall of Fame—driver survived.

thirty-five million dollars a season for each entry, which can require as many as sixteen tube-frame chassis. Why so many? Because while the chassis may look identical, there are slight variations in each one depending on the track on which they're designed to run. Engine costs—and many teams lease their engines—can run anywhere from three to five million dollars per car for a season.

Of course there are countless other cost considerations involved with getting a race car out onto a NASCAR track to compete, but this just begins to scratch the surface. How on earth can a sport as homegrown and humble as stock car racing, with roots that wrap all the way back to bootlegging days, attract that kind of money and make its power players some of the wealthiest in any sport? How can something that looks as boring as it can on television (I'm sorry, but if you don't understand what you're watching, it can be boring to watch) pull in and hang onto that many fans, often for generations? These are questions that many at NASCAR are asking themselves all over again, as the sport is looking to almost reinvent itself and attract a wider audience, to cultivate a younger generation of fans. As some in NASCAR have told me, that's no small undertaking. Young people today have a much shorter attention span than their counterparts of a generation ago. They have cell phones. They have iPads and Nooks and Kindles. They have Facebook (though my children have told me that we old folks have scared away the younger crowd), Snapchat, Vines, and Twitter. For those of you out there who have tweens and teens, when is the last time you saw one of them devote an entire day to any type of entertainment that they couldn't hold in the palm of their hand?

I'll admit that at first, I myself was reluctant to tackle this challenge of learning all I could about NASCAR in a single season, but I have to say that now that I'm on this side of it, I'm very glad I did. NASCAR is a success story that's made up of people from all walks of life, a real American success story. Its history is captivating to me. The fact that stock car racing's roots are intertwined with Prohibition and bootlegging is a connection that for some reason or another I find deliciously fascinating. Dare I say it, there's something bad-boyish and kind of sexy about all that. These juicy details, the domination and success in motorsports, all these fascinating tidbits about NASCAR were too much for me, a news reporter, to ignore. I had to know more.

Chapter Two

Packing to Travel Outside My Comfort Zone

How have I missed out on the cultural phenomenon called "NASCAR" my entire life? Oh sure, I've known NASCAR fans. I've blown right past glimpses of the Sunday races as I clicked through the two hundred or so TV channels served up daily by our family's satellite dish. I've heard race car drivers being interviewed on the evening news every now and then. All I knew about NASCAR, however, was that for most of my life, I had no interest in it at all. I can freely admit that, now that those dark days of ignorance are behind me. For a variety of reasons, not the least of which includes the fact that I couldn't stand any longer being in the dark on such a hot topic, I decided to pack my bags and embark on a nearly yearlong adventure, visiting cities throughout the southeast that are home to legendary racetracks such as Daytona International Speedway (nicknamed "The World Center of Racing"), Bristol Motor Speedway ("Thunder Valley"), Talladega Superspeedway ("The Big One"), Darlington Raceway ("The Lady in Black"), and of course, Atlanta Motor Speedway ("Real Racing, Real Fast"). By talking to locals, fans, media pros, and some members of racing royalty, I hoped to understand the pull, the attraction, to stock car racing. What had previously seemed to me to be an endless, mind-numbing run of beer drinking, left turns, horrific crashes, and hours of roaring noise might actually make sense if I talked to those closest to the frenzy.

That's exactly what I've done here, and I'm a better woman for it. In addition, in the interest of full disclosure, I had my own selfish reasons for wanting to learn more about NASCAR. I finally decided I had heard quite enough about being NASCAR ignorant from friends, colleagues, and even family members. My curiosity about this sporting and cultural oddity called "NASCAR" simply got the best of me.

When I began this journey in early 2013, NASCAR was something that was situated very far outside my comfort zone. I had become complacent in my suburban cocoon, very comfortable in what I'm grateful to say has thus far been a contented, safe, even-keeled life. Stepping outside my comfort zone does not come easy to me at the settled age of fifty-two. I detest how that sounds, and I wish it weren't true, but it is. I must say that since my husband and I became empty nesters a couple of years ago, I haven't quite known what to do with myself and the extra time that walks in the door when the last child walks out. That fish-out-of-water feeling, combined with a reporter's curiosity and the void I've been trying to whistle past now that our children are grown, gave me just enough of a kick in the pants to propel me into this strange new world I mistakenly believed to be the sovereign territory of rednecks and motor heads.

I began my research by reading, lots and lots of reading, because reading falls well within the boundaries of my ever-important comfort zone. Intrigued by the information I uncovered about NASCAR and stock car racing in general, I then got my hands on film of some famous past races. I watched the 2001 race that ended "The Intimidator" Dale Earnhardt's life. That Sunday, among NASCAR fans anyway, will forever be known as Black Sunday. I watched the 1988 Daytona 500, which was the first and only race to have a father and son (Bobby and Davey Allison) win first and second place. I also watched the 1987 Talladega race in which Bobby Allison was in a horrific crash, an event that led to the eventual use of restrictor plates under the hood, devices that rein in some of the horsepower that those powerful cars can churn out on such a long track. While I understood very little about the intricacies that are documented forever in those old race films, what I did see was excitement, danger, and the incredible skill of both the drivers and the mechanics in the pits. I saw sold-out grandstands. I saw passion. I saw a phenomenon.

My timorous initial research done by reading and watching films underscored the fact that I was just dipping my toes into the racing waters, because I saw early on that NASCAR is an experience, not merely a topic. It became quite evident to me that in order to experience NASCAR, I'd have to do a whole lot more than simply read about it and watch past races. I'd have to venture out of my quiet, safe, climate-controlled life and go to where the action really is. I had to go to a race, and it only made sense to me to begin with the big one that kicks off every NASCAR season. I made up my mind to buy a ticket and make the trip to watch the Daytona 500, to see the race live and in person.

The prospect of going to something as foreign to me as a NASCAR race—by myself, no less—made me very uncomfortable. Nonetheless, my husband had a carved-in-stone business commitment that he simply could not reschedule, and my children were scattered at college, work, and other young-adult-life commitments. After asking two of my girlfriends to go with me and getting a resounding "No!" from both of them, my choices were either to go to Daytona by myself or not go at all. Looking back, I think I knew even then that not going would have been a mistake, that missing that first race was simply not an option if I really meant to unwrap the phenomenon of NASCAR. So I packed my bags, my backbone, my reporter's skills, and my undying adoration of the region in which I've lived most of my life, and I hit the road. I may as well have packed my suitcase to visit Mars, because I had no idea what I was getting myself into.

Chapter Three

Traveling in the South
Never Disappoints

The prospect of traveling throughout the South is a fine one. Since it's a well-known fact that NASCAR started in the South, and since the South boasts some pretty famous and exciting racetracks, I couldn't think of a better region of the country in which to travel for my research. I imagine there are worse ways to spend the better part of a year than tooling around Southern back roads, going from one town to the next, and soaking up all the entertainment that I anticipated along the way. I've traveled much of the United States and a few foreign countries in my lifetime, and I have yet to visit any place that offers as many elegant, historical, funny, quirky, and just downright weird surprises as the southern United States. That may sound like a boastful claim to make since I live here, but if you've ever had the opportunity to spend some time here, you know just what I mean. The South graciously serves up stately old mansions, grand live oak trees draped with flowing Spanish moss, and towering, glossy magnolia trees dotted with their telltale gaudy white blossoms, regardless of whether you are traveling winding country roads or eight-lane interstates. You've eaten in fine restaurants as well as converted gas station dives that serve fried catfish and hushpuppies. Without a doubt, they both had "sweet iced tea" listed on the menu, just as surely as they had salt and pepper on the table. You've enjoyed the spectacle of fine modern art displayed in bustling, diverse cities, and you've seen hillside trailers with motorcycles and washing machines sitting side-by-side on the wobbly front porch, a tattered Confederate flag waving defiantly over the whole mess. This region of the country is a wonderland of glorious contradictions.

The southern United States is beautiful, with its lush, rolling hills, evergreen palette splashed with impossibly brilliant flowers, picture-postcard pastoral scenes, and wide open sky. The cities are alive and vibrant, every bit as complicated as New York, Washington, DC, and those places, but there's that something extra down here that you simply won't find anywhere else. There's a red carpet of hospitality here, and yes, even though we have our share of transplants who've moved down from up north, that flavor of warm hospitality still lingers. Oh, and the accents. My heavens, do I love a Southern accent. Listening to a true Southerner speak the Queen's English is very much like listening to a lilting melody, hypnotic and entrancing. Perhaps that's why Southerners make such good storytellers.

The weather here is, for the most part, pleasant and agreeable. Everywhere in the South, with the exception of the southernmost reaches of Florida and perhaps Louisiana, has four distinct seasons. It's oppressively hot and humid in the summertime, and winter is a mishmash grab bag of heat, snow, ice, and even tornadoes. Spring and fall last about fifteen minutes each, but what a glorious fifteen minutes those are! I love the long, hot days that seem to go on forever. I love the fact that Old Man Winter rushes in, gives us a swift kiss on the cheek, then rushes off again, only to return about eleven months later. I love the prim and proper coastal towns of Charleston and Savannah; I love the farmland of south Georgia, Alabama, and Tennessee. I've lived in New York and a few other places in my lifetime, but I always came back home to Georgia. I suppose I'll die here someday, and that's OK with me.

Going to the races that I did, visiting the NASCAR Hall of Fame and the big, impressive motorsports garages, all required road trips here and there, and that meant adventure and surprise. In the

South, you can be driving along a serene, winding country road, flanked on both sides by cattle contentedly grazing or horses kicking up their heels on an early spring day. Around the next curve, you can come face-to-face with a forty-foot-tall statue of Jesus wearing cowboy boots and eating a corndog, the entire structure made of motorcycle parts and old carburetors. That's just how we are down here: charmingly, strangely unpredictable.

Somewhat early on in my research, I decided that touring the NASCAR Hall of Fame and a couple of the shops that build race cars for NASCAR drivers was an absolute must. My husband and I set aside a long weekend for the trip to Charlotte and Mooresville, two towns in North Carolina located conveniently close to one another. We hadn't been on the road very long when we started coming across the very things I've been talking about, and it all began with bizarre billboards. As soon as we crossed the Georgia-South Carolina line, we spotted a mammoth billboard advertising "Legal Palmetto Moonshine," the closest thing to the illegal stuff you can get these days, so the billboard claimed. The still, boasted the billboard, has real copper coils, and the moonshine is bottled in clear jars with metal lids. "It's the Real Deal," shouted the ad. Nailed to one of the posts supporting this billboard advertising hooch was a smaller, less eye-catching, hand-lettered sign telling travelers to be sure to visit Beaverbrook Baptist Church. Everywhere you look in the South, there are contradictions and humorous spectacles to ponder. In fact, I compiled a list of some of my favorite signs and messages along back roads throughout the South, all chucklers for one reason or another. We saw signs for "Flippin Church of Christ" (in Flippin, Kentucky), "Hot Fresh Boilded P-Nuts," (a misspelled but delectable southern delicacy) in Alabama, "Ho-Sale Pillows" (sold out of a blue school bus parked on the side of a road in south Georgia, by a woman who seemed to be quite virtuous), and my personal all-time favorite billboard advertising home-style country cooking: "There's plenty of room for all God's creatures, under the gravy."

Billboards are a cultural oddity, an interesting form of equal parts marketing, artistic expression, and spiritual commentary down here in the South, and I'll tell you why. I am convinced that God himself has invested in a string of the things stretching from Florida all the way up to Virginia. That might sound like a strange thing to say, blasphemous even, but hear me out. As Brother Bill, the chaplain at Atlanta Motor Speedway whom you'll meet later on, reminded me: "This is the Bible Belt." I lost count of the number of billboards we saw on the way to North Carolina that Thursday, each of them sporting messages like, Don't Make Me Come Down There – GOD, and There Are No Atheists in Hell – GOD, and God Died for Your Sins, So Slow Down! Now I don't know about you, but I'm not about to put anything up on a billboard, sign God's name to it, and think there won't be any repercussions. Who else could it be, sending those cryptic messages?

Here's another missive at which I laugh aloud every time I read it, and believe it or not, it appears often throughout the South (well, in Alabama and South Carolina, anyway): Brush Your Teeth For Two Minutes, Two Times a Day. The only picture on the billboard is a toothbrush casually plopped into a rinse cup. Really? Now as I said, I've only seen these billboards in Alabama and South Carolina, so deduce from that what you will.

One of my favorite sights along any back road in the South is a local fruit stand. A true local stand is one that offers fruits and veggies grown by one or a few local farmers. Some of the best fresh fruit I've ever had the pleasure of tasting came from one of those little roadside peddlers. Many offer for sale the South's answer to caviar that I mentioned earlier: boiled peanuts. They sound awful, I know, but they really are delicious. Some of the bigger fruit stands in South Carolina also offer fireworks for sale, since it's illegal in Georgia to buy them.

More often than not, the signs directing travelers to stop and check out the luscious local produce are hand painted. Signs hammered to stakes and driven into the ground at rakish angles read something like this: Boiled P-Nuts. Cantalop. Watermellon. Maters. And on, and on. Not one of those farmers ever spells out the word "peanuts," and no one who makes those signs has any idea how to spell "cantaloupe," "watermelon" or "tomatoes." Of course, neither do some of our nation's past presidents and vice presidents, so that's neither here nor there, I suppose. Still, if you're ever down here for a visit, be sure to stop and buy a basket or two of fresh fruit from one of those roadside stands. Pick up a watermelon in late June or early July. Load up a basket of crisp fall apples in October. None of that genetically altered, picked-too-early fruit you find in the big supermarket chains will ever taste good to you again.

Speaking of fruit, there's a little town in South Carolina called Gaffney. As soon as you approach this town on Interstate 85 north, you can't help seeing their four-story water tower right alongside the highway. It's a peach. It's a 150-foot-tall peach. Whoever built it did a fantastic job, and when we drove by it that day on the way to North Carolina, I must say, it looked strikingly like a giant, fresh-picked peach. Not everyone is aware of this, but even though Georgia is known as "the peach state," South Carolina actually produces more of the tangy-sweet fruit than her southern neighbor. Anyway, the last time I saw Gaffney's peach before this trip was about twenty years ago. It was in dire need of a paint job, as it had faded to a pale, sickly, fleshy pink instead of the vibrant peach color it should have been. In truth, the thing looked like a four-story human fanny up there on a pedestal, obscene really. I was very glad to see that the city leaders had seen fit to do something about it.

That highway and back road trip, which I think took us about four or five hours, was simply divine. We really enjoyed ourselves that day, stopping for lunch at a place called Bluegrass Heaven Barbecue, with a dine-in area about the same size as the window unit air conditioner blasting cold air in the back corner. The menu was limited, but the barbecue was fabulous. The breakfast menu boasted bologna biscuits, which we'd never heard of before. On the way out, we picked up a couple of really fine cigars and a baseball hat that had HAD A PEACH OF A TIME IN GAFFNEY printed on it. You just never know what you'll find in these places. But let's get back to talking about NASCAR.

Chapter Four

A Little Bit of Racing History

Even those who have just a passing familiarity with NASCAR have surely heard the tales of its origins in bootlegging, and those tales, as it turns out, are true. Today's high-tech, sleek, and powerful automobiles are in fact the direct descendants of bootleggers' souped-up cars, which were of course created out of necessity born of Prohibition. Junior Johnson, one of the first inductees into NASCAR's Hall of Fame, credits his driving skills to his early days of running liquor and outrunning the revenuers. But the moonshine-soaked roots of stock car racing don't stop with just the drivers; some of the most skilled mechanics ever to turn a wrench in the pit cut their teeth on cars used to transport corn squeezings. When those first drivers made the switch to racing on tracks rather than surreptitiously navigating back country roads, they brought their mechanics with them to the pits. Who better to keep their Chevys, Fords, and other stock cars in racing shape? And by the way, the cars are referred to as "stock" cars because they look very much like the regular stock that car manufacturers turn out year after year, right up until the hood is raised. What lies underneath the stock car's hood is a marvel of mechanical and engineering brilliance, the product of millions of dollars of research and decades of experience. When I first laid my eyes on the polished engine that's housed under the hood of a

NASCAR Car of Tomorrow, my first thought was that I was looking at the shiny chrome muscled gut of the machine. It's an impressive sight.

It wasn't just the early drivers and mechanics who could list liquor law violations on their resumes; race promoters and track owners had ties to bootlegging, as well. The stories, which are officially depicted right there for all to see at the NASCAR Hall of Fame, tell of an infamous, dirt-poor bunch of country boys who lived anywhere from Georgia to Alabama, Tennessee to Virginia. Junior Johnson's actual moonshine still is on display in the Hall of Fame. While Junior spent a year in an Ohio jail for owning and operating a still (so did his daddy), he never got caught actually transporting the bootleg whiskey. Maybe that's why Johnson, the driver credited for discovering a bold and sometimes dangerous racing maneuver called "drafting," won fifty NASCAR races in his career.

When whiskey was labeled "illegal" by the US government during the thirteen years of Prohibition, these good ol' boys had to find a way to survive, and they did so by running illegal whiskey from Point A to Point B. They muscled up their cars for the simple reason that they had to outrun the law, who was always chasing them. Back then, they were racing to survive. Of course not all of the early drivers were bootleggers. Some were just good ol' boys who loved racing and were looking to do it on any dirt track they could find, usually on Sunday afternoons. Mr. Ed Clark, the president and general manager of Atlanta Motor Speedway in Hampton, Georgia, put it to me like this: "We've all known guys like that. They were just a little bit crazy, leaned to the wild side a bit and always seemed to find a way to go a little faster, to cheat death a little bit closer. Those guys were early stock car drivers too."

NASCAR's connection to bootlegging has not always been considered acceptable, or even historically rich or quaintly amusing. In fact when World War II ended in 1945, quite the opposite was true. A celebratory stock car race was planned at Lakewood Speedway in Atlanta (a perilous dirt track built around a lake and originally used for horse racing) to jubilantly mark the end of the

long war, but there was just one problem. There were five drivers
who threw their hats in the ring to race at Lakewood that day, and
every one of them had a record of bootlegging to their credit. Well,
some of the fine upstanding citizens of the state of Georgia were
having none of it (even though their daddies or their own brothers may very well have run a quart or two of homemade booze
in their day), and in indignation they insisted that the entire event
be cancelled. Editor Ralph McGill of *The Atlanta Constitution* (a
conservative newspaper in those days) jumped on the bandwagon
too, running several editorials protesting the sheer scandalousness
of the very idea of bootleggers sullying the soil at Lakewood. Some
credit McGill's editorials with running many of the most popular,
if somewhat shady, drivers out of Atlanta and up to North Carolina
after World War II, where stock car racing got a strong foothold
in the famous clay tracks of the hill country. Newspapers carried
headlines of Atlanta mayor William B. Hartsfield himself showing
up on race day, backed up by law enforcement officers prepared to
take the lawbreakers away in handcuffs as soon as they showed up
to race.

All this ruckus ended in three of the five outlaw drivers being
banned from the race by the nervous promoter, who wanted no
part of a legal showdown or bad publicity in Atlanta. The other two
drivers, who had gotten wind of what would be waiting for them
if they made an appearance at Lakewood Speedway, didn't even
bother to show up on race day. What the jittery promoter didn't
foresee, however, was the angry reaction from the thirty thousand
ticket holders for that race, who were not at all happy that they'd be
cheated out of the chance to see former north Georgia moonshiner
and wildly popular driver "Reckless" Roy Hall race that day.
Hall, incidentally, hailed from the same Georgia town as famed
NASCAR driver Bill Elliott ("Awesome Bill from Dawsonville"),
who added more affable Southern spice to the sport beginning in
the mid-1980s. Elliott set the fastest qualifying speeds at both the
Talladega and Daytona speedways by topping 210 and 212 mph,
respectively, and he won "NASCAR's Most Popular Driver" award
a record sixteen times. When he was nominated a seventeenth time,

he graciously bowed out of the competition. Must be something in the water up there in Dawsonville. Faced with tens of thousands of disgruntled fans, the promoter flip-flopped again and at the last minute lifted the ban on the five drivers. The rest, as they say, is history. "Reckless" Roy blew the doors off the other drivers on the track that day, winning the race and stirring up a hornet's nest of outrage to boot. That race at Lakewood Speedway, combined with the eventual across-the-board ban of drivers with "hauling " on their records, was just too risky for many of the race promoters of the day, and they bailed out of the business as a result. Another legend tells of stock car driver and Alabama moonshiner Bob Flock showing up at Lakewood for a 1947 race, intent on driving under an assumed name to avoid arrest. As he drove onto the track to take the green flag, federal agents recognized him, then chased him for a lap or two before Flock drove right through the fence, outrunning the cops until he eventually ran out of gas. Telling the story years later, Flock reportedly said, "I would have won that race if the cops had stayed out of it." All things come together for the good of those who love racing, as they say in the sport, and the scattering of nervous promoters made way for the granddaddy of all race promoters to enter and take center stage. Bill France, or "Big Bill," as he came to affectionately be known, was back then a gas station owner and relatively new fringe race promoter, with his innovative ideas and long-term vision for stock car racing. He was a driver as well, and in 1936 he raced and finished fifth in the inaugural stock car race held on the sandy Daytona Beach Road Course. Understanding the sport from both sides as Big Bill did, he knew that these Sunday afternoon races being run all throughout the South were poorly organized, wildly dangerous events in which drivers often had to chase down promoters just to get paid. France knew some of the banned outlaw drivers; many of them were trusted friends. He knew stock car racing and already had a good thing going in Daytona. The ban on outlaw drivers had left a gaping hole in racing, and France was astute enough to know that. Naturally, bootleggers or not, drivers still raced each other whenever they got the chance, but France had the savvy and connections to pull all the elements and players together. That December 1947 meeting opened doors and paved the way for the official establishment of NASCAR on February 21 of the following year. To this day, the France family still owns NASCAR outright and maintains a controlling interest in racetrack operator International Speedway Corporation. Bill France's son, Bill Jr., served as the head of NASCAR from 1972 to 2000, and his son Brian took the helm in 2003 as chairman and CEO. Lesa France Kennedy, the daughter of Bill France Jr., has been named "The Most Powerful Woman in Sports" by *Forbes*, and serves as CEO and vice chairperson of the board of directors for International Speedway Corporation. She also serves as a vice chairperson of NASCAR. In 2006, Brian France was recognized by *Time* magazine in the annual "Time 100: The People Who Shape Our World" issue. A savvy and philanthropic businessman, France has surrounded himself with brilliant business people who love the sport. He has grown NASCAR, a Southeast-based sport since its inception, into an international phenomenon and the most heavily licensed, brand-respected corporation in the world. The history of NASCAR and a discussion of its power brokers can't be fully addressed without giving a sincerely respectful nod to Bruton Smith, the eighty-something-year-old NASCAR heavyweight and North Carolina–born promoter, owner, and CEO of Speedway Motorsports, Inc. (he's a NASCAR track owner). The owner of eight NASCAR tracks, including Atlanta Motor Speedway, Smith is unquestionably a top NASCAR figure and shrewd billionaire businessman. He was inducted into the International Motorsports Hall of Fame in 2007 and the North Carolina Business Hall of Fame the year before. He was nominated for the NASCAR Hall of Fame in 2013, a move that many in the sport have said is long overdue. It's no secret that he and the Frances have butted heads and even landed in court over various racing-related issues, including venue closings supposedly designed to drive attendance to other tracks, and suggested conflicts of interest. This turn of events is just another example of the true passion that exists within NASCAR, from the apex where the Frances and Smith reside, to the drivers themselves, who feud openly and often. That feuding takes place right there on the track in the form of a fight after a race (picture a baseball pitcher throwing at a batter, and the dugouts from both teams emptying onto the field). No one really throws a punch hard enough to do any damage, but it's the thought that counts, a heated exchange of words. Today racing rivals even feud on Twitter, a fact that I find a telling and amusing nod to today's instant-gratification communications. Never let it be said that NASCAR hasn't kept up with the times.

There's been a matter of contention about NASCAR's geographical roots through the years, and that's the question of whether those roots lie in Charlotte or in the cities of Atlanta and Daytona combined. There's no doubt that stock car racing's roots run all along dirt tracks throughout the South. For years, North Carolinians have claimed bragging rights to being the epicenter of stock car racing, and in 2011, stock car racing was named the state sport. In 2006, the new NASCAR Hall of Fame was built in Charlotte. Add to that the fact that most NASCAR teams with their high-tech, multi-million dollar garages are based in and around

Exterior of NASCAR Hall of Fame, Charlotte, North Carolina.

Charlotte. Going even further back in stock car racing history, some say that when the sticky red clay of the North Carolina Piedmont is mixed, graded, tamped, and watered in precisely the right manner, the result is an even, perfect racing surface. Many a race has been run in a farmer's field in the days before the organization and structure of NASCAR. One must consider, too, that North Carolina is always very well represented at NASCAR races, with drivers like "King" Richard Petty, "The Intimidator" Dale Earnhardt, "The Last American Hero" Junior Johnson, and many more talented racers hailing from the twelfth state in the Union. All this being said, If I had to pick a state, I'd go with North Carolina as being the hub of the sport. Nevertheless, it's important to keep in mind that the first NASCAR meeting was held at the Streamline Hotel in Daytona Beach, Florida, and Bill France Sr. had strong ties to that area. NASCAR headquarters is located in this Florida beach town. Daytona definitely stands out as a key NASCAR city, considering the fact that the Daytona 500 kicks off every NASCAR season. Atlanta, of course, boasted Lakewood Speedway early on, a dirt track originally built for horse racing. It was a very dangerous track for stock car racing, as several cars ended up in the lake instead of crossing the finish line. Racing at Lakewood stopped in 1979.

Richard Petty's #43 car displayed in Hall of Fame.

There was the legendary quarter-mile track in the heart of Atlanta known as the Peach Bowl Speedway. Host to many varieties of racing including NASCAR convertibles, stock cars, jalopies, and such, the Peach Bowl was sold, then closed, in 1970. I'm told that land is now home to an Atlanta rapid transit bus repair depot. But in 1960 came Atlanta Motor Speedway, one of the big, shiny, new-fangled tracks dubbed by some experts the fastest track in racing, built just south of Atlanta. As the undisputed Queen City of the South, Atlanta has been in the stock car racing landscape since the sport began drawing sizeable crowds and hefty purses. Frankly, while purists may argue these fine points about cities and firsts, I don't think it really matters in the overall scheme of things. What matters is that stock car racing sprouted in the South, and that's where a lot of the personality and color of the sport and the players come from. Brilliant public relations and marketing genius churn and generate the money in NASCAR, but ultimately it all boils down to family, fun, and excitement. Stock car racing is a sport (though some will argue that it is not; I'll let you decide for yourself as we go along) loved by fans across the entire nation, even way out in California where electric cars and "green" thinking are all the rage. It's loved in other countries, just as American-made films, music, and other trends are loved. It's a spectacle that's bigger than life. The cars that race in NASCAR competition are powerful, fuel-guzzling, heavily muscled products of engineering genius. The male and female mechanics who work their magic on these cars have specific training on how to do so. There are mechanical and aerospace engineers on racing teams, and there are also the guys who are just very good at turning wrenches and tweaking engines. We Southerners love that stuff. NASCAR is steeped in pride, tradition, and legacy (we love that stuff too); it's not at all politically correct, and it involves strength, speed, and intelligence (all powerful aphrodisiacs if you ask me, but let's stay on topic). The legends of the sport are often marked with profound tragedy, with fathers and sons losing their lives doing what they love best. Tragedy often turns up a hero or two, and NASCAR surely has its share of heroes. For instance, both of driver Bobby Allison's sons, Davey and Clifford, died within a year of each other. In 1992, Clifford was killed during a practice-run crash for the NASCAR Busch Series race at Michigan International Speedway. The next year, Davey was killed in a helicopter crash when he tried to land his copter in the infield at Talladega Superspeedway. In 1988, Bobby and Davey finished first and second, respectively, in the Daytona 500, a feat never accomplished before or since. But Bobby's near-fatal, career-ending crash just four months later during the opening lap of the 1988 Miller High Life 500 at Pocono Raceway caused major head injuries and profound amnesia in the patriarch. I have read that today, Bobby Allison does not even remember that once-in-a-lifetime achievement with his son at Daytona, though the hope among fans is that someday, he will. Many fans, no matter who their favorite driver may be, talk about that tragic year for the Allison family with tears in their eyes, just one of many testaments to their dedication to and respect for the drivers.

Largely considered a blue-collar spectator sport (and therefore inexorably linked with the South), NASCAR is so much more than that. Incidentally, judging by the displays of lavishness and material comfort I saw on parade at each and every race I attended, I'd have to say that the fan base includes both blue- and starkly white-collared fans, as well as more than its share of celebrity fans. That hasn't always been the case, though. How did NASCAR grow from its early days in the dirt with self-proclaimed rednecks its primary fans to its current appeal to a diverse audience that includes "yuppies" (a favorite fan term) and celebrities, as well? It has something for everyone, that's how. I was about to get my first taste of just how true that statement is, because for the first time in my life I was headed to a NASCAR race, to experience the legend of Daytona for myself, firsthand.

Chapter Five
Off to My First Race

It was February, and there I was, just one in a sea of more than a hundred thousand zealous fans, clutching my access pass in my sweaty palm. I was grinning and tapping my feet, my ears numb but giddily ringing, hanging on to the last notes of the music the fans had just been treated to. The Zac Brown Band had just wrapped up a fabulous concert for the crowd, a little prerace treat that had whipped the masses into a good-natured frenzy for the main event. The fans were happy, they were primed, and they were very, very loud. I made a few mental notes to myself as I followed their lead and stood up again, waving my checkered flag and cheering at the top of my lungs. *Next time,* I thought, *I'll be better prepared.* I suppose right then and there was the first time along this journey that I had made up my mind that there would even *be* a next time. I had taken my sweet time on the drive to Florida, stopping along the way from Atlanta to check out the Allman Brothers Band Museum (like stepping back in time to the foggy 1970s) and the Georgia Sports Hall of Fame in Macon, which I must say I found to be a bit boring. In all fairness, I was bored with it because of my own lack of sports knowledge; I know a little bit about several sports, but not a lot about any of them. I was dragging my feet a bit on the 450-mile drive to my ultimate destination, headed to a place and an event that was so far outside my comfort zone that yes, I do believe I was procrastinating. I finally arrived in Daytona Beach on Saturday, and the next day, there I stood among a crowd of more than one hundred thousand fans, praying that I didn't stick out like too much of a sore thumb in my crisp white linen suit and high heels. Nicole Kidman pulled that look off nicely as Tom Cruise's girlfriend in the movie *Days of Thunder*, so I thought I'd give it a try. I was already regretting it. There's a lot of walking in NASCAR.

I was mentally pinching myself; I had overwhelming trepidation and nagging apprehensions about making this trip (which were already fading and seeming downright silly to me), but I had bought my ticket and gone anyway. All the research, waiting, planning, worrying, and doubting had led up to this moment. I believe I was every bit as excited about having made this trip by myself as I was about actually being there. At 1:00 p.m. precisely, the noise level at Daytona International Speedway ratcheted up to an intensity I've never experienced in my entire life. I thought the concert and prerace festivities were loud, but this was thunderous, booming. At exactly 1:00, announcer James Franco's commanding voice boomed through the loudspeaker, "Drivers and Danica, start your engines." The grandstands shook. The fans went wild, and as soon as Franco uttered those last words, the rich, rolling smells of oil, gasoline, and rubber filled my nostrils. I felt lightheaded and exhilarated at the same time. The thick smells also made me feel a bit nauseated, but I quickly got used to them and hardly even noticed after a while. My heart was thrumming and vibrating so much that I feared it might jitter right out of my chest. I turned to sit back down after the announcement, and just as I was wedging my derriere into my cramped seat, I noticed that my sweaty, deliriously happy neighbors on both sides were still standing, so I followed suit. No one else was sitting, so neither would I. *Blend*, I thought. *Blend!* I shielded my face from the sun with my hand, my eyes squinting but unblinking. I fought valiantly to resist the urge to cover my ears in order to muffle the growling roar coming from the cars whose drivers had just awakened them. Merely covering my ears would have had little impact on the sound anyway.

Driver Danica Patrick drives behind the pace car before race gets into full swing.

My senses were on maximum overload, zinging and pinging all over the place, especially once those mammoth car engines roared to life. I have to confess, there's something powerfully exciting, sexy even, about the rumble of forty-three muscle cars firing up their super-engines at once. It's exciting when those cars start moving, almost slowly at first as they come off the starting position, warming up their tires and beginning to get a feel for the track. Then that first lap behind the pace car is complete, and the drivers open up a little more, then a little more, and then the green flag waves, and then they're wide open, and the crowd roars in frenzied excitement. Then they're flying around the track, jockeying and drafting with breathtaking skill. I had to focus on my rate of breathing in order to avoid hyperventilating. I know how ridiculous that must sound, but I am a quiet woman who treasures peace and tranquility. The noise was, in a word, tremendous. The smells were nearly overpowering, and the speed was both alarming and exciting.

There was a woman sitting across from me, all the way across the track, who was probably in her mid-fifties—my neck of the aging woods. She stood out to me even from that distance, because she looked like a very energized patriotic dot from where I sat. Fortunately, I had remembered to take my binoculars with me to the race. I had learned in my early research that a true NASCAR fan never leaves home without her binoculars. My personal addition to that wisdom is that a good pair is small enough to fit into a stylish purse but powerful enough to zoom in on the minute details of a fiery crash, a drunken brawl, or an overzealous woman whose halter top was tied too loosely when she left the camper that morning. The binoculars for which I shopped extensively and eventually purchased are official Dale Earnhardt Jr. high-power compact binoculars, and they most definitely do the job. Yes, field glasses made my list of "must-haves" months ago, and they stay tightly wedged in what I've come to call my "race day bag," along with permanent markers, some hidden cash, SPF 30 lip gloss, hand sanitizer, pepper spray, and sunscreen.

I couldn't help zeroing in on my excited friend "the dot" across the way with my official, licensed binoculars. She was very well tanned and oiled (and it was only February!), wearing a red, white, and blue bikini top and cut-off denim shorts. I self-consciously compared her attire with my own that day, that impractical white linen suit and torturous heels. I looked more appropriately dressed for the Kentucky Derby than the Daytona 500, but that was another lesson learned. My comfortably clad friend across the way was jumping, shouting, jiggling, and overall just very happy to be there. I certainly understood her enthusiasm. I felt as though I'd been beamed up from earth and plopped right smack down in the middle of a very different planet. As a ticket holder that day for the fifty-fifth running of the Daytona 500 at the famed Daytona International Speedway, a high-banking track built in 1958 by Bill France Sr., I was rubbing elbows with about 150,000 thrilled fans (I've read that the Speedway seats 168,000 people). I was attending the first NASCAR race of my life, and I was completely overwhelmed. But what better way to learn and eventually understand stock car racing?

The acrid smell of hot rubber assailed my nostrils, making my eyes water and sting. The hum and roar of the tires on the asphalt once the cars really got going were relentless; I remember thinking my head might actually split in half like a ripe melon that's been laid open for mid summer consumption, only no one around me would hear it or even notice it because there was a race going on down there.

I noticed many people in the stands wearing headphones, and finally I mustered the courage to ask the woman sitting next to me where I might find some of those wonderful headphones. I didn't tell her this, but I would have happily traded my car for a pair if necessary. Three hours of that noise, give or take a few minutes? I couldn't have stood it without some headphones. The woman's name was Janice, and she kindly directed me to the vendor area, where I was more than happy to part with the cost of renting ear protection. That was how, at my very first race, I learned that there is a wonderful alternative to enduring the relentless buzz of forty-three unmuffled V-8 engines, and that is the use of scanner headphones. The cost to rent them for the day was about sixty dollars, and in hindsight that was the best use of sixty dollars I can remember in, well, forever. What I didn't know was that those handy headphones came with a scanner (which the guy at the counter happily handed me and began to demonstrate, but I kept shouting, "No thanks, that's OK, just the headphones!"). He finally gave up trying to show me how to use the scanner, but by taking the man's advice and tuning into certain channels, I could have actually listened in on the back-and-forth chatter between drivers, their pit crew chiefs, and spotters up in the stands. Listening to all that wouldn't have done me much good at that first race anyway, as I would have had no idea what I was listening to. Listening to it during the last race of the season, however, would have been very interesting. A lot goes on over those radio frequencies, as NASCAR fans worldwide would learn when the 2013 season drew to a close. Some exchanges between drivers and team members went out on those airwaves and into the headsets of fans and officials alike at that last race in Richmond, and the repercussions were felt all the way from the Chase for the Cup to the NASCAR rule book.

Incidentally, for those of you new fans who may be considering attending a NASCAR race for the first time, let me just warn you up front that the headphones are not at all flattering. They don't do much for your hair, either, but neither do the sweat and humidity you're sure to endure once summer in the South heats up. Just make up your mind before you go that race day will not be a good hair day. I toughed it out though, and have since come up with a few tricks (visors and scarves, mainly) to get me over that hurdle. I'll also add that by the end of the season, I was having so much fun that I didn't care in the least what my hair looked like during a race. That might sound like a trivial victory to some of you, but if you knew me, you'd understand just how much of a breakthrough that was.

Janice and her husband, Todd, were great neighbors to have at my first NASCAR race. They lived up to what other fans, mechanics, track managers, and even some drivers had told me: the fans are great, and if they know you're new, they'll help you understand the race and what's going on as much as possible. Janice had wildly curly red hair and thousands of freckles on her friendly face. Her nose and the tops of her ears were covered with the telltale thick, white paste of Zinc Oxide, which she offered occasionally to fans sitting near us in the stands. She offered some to me, but I politely declined the gesture. I was already wearing enough white, and I really didn't think I had to worry about getting too much sun. As it turns out, I was wrong. Even with puffy clouds filtering the February sun in Daytona, using sunscreen is a good idea. I walked out of that stadium looking like a soot-covered, medium-rare steak.

Todd was just as quiet as Janice was gregarious, and he was very intent on the infield and pit crew activity leading up to the race. Although I didn't know either of them when I arrived, we became quite friendly by the end of the day, even exchanging phone numbers before we left the track later. I had assumed the people who had headphones that day had brought them in when they came—and some had—but thank goodness she told me it wasn't too late to rent some. She may very well have saved my hearing. Janice, if you're reading this, thank you. When we first began talking, she had been excitedly chattering on about something or another for a good fifteen minutes. I finally figured out that she was talking to me. I had no idea what she was saying, but she was incredibly animated about whatever it was. She could have been telling me about the cute overpriced #18 T-shirt and black-and-white-checkered baby bib she had just bought her grandson, or she could have been telling me that my hair was on fire. I had no way of telling. My ears were actually numb. Still, I was fascinated by my surroundings and didn't want to blink for fear I might miss something. After I rented my headphones, Janice and I were able to carry on conversation in short, shouting spurts, though we did most of our communicating via text messaging. She and her husband were longtime race fans, diehard followers of famed driver Kyle Busch, #18 (to both the drivers and the fans, the numbers displayed on the top and sides of the cars are every bit as important as the name given the drivers at birth). They wore Kyle Busch gear from head to toe, and the camper they pulled behind them from Birmingham to Daytona had "#18 Kyle Busch" painted on both sides. They had taken an entire week of vacation to make this trip. Janice and Todd took turns trying to school me in racing basics during the three-plus hours we sat together. In fact, Janice is also the one who advised me against waving my flag and foam finger too often during the race. I think she feared for my safety, as she should have—I would not recommend waving these items and blocking the view of those sitting behind you at a race. I added a few words to my vocabulary list that day, donated by some of the folks whose views I occasionally impeded. The finger and flag, after some coaxing, also fit nicely into my bag.

Everywhere I looked, there were marvels I don't believe I'd have ever seen had I not landed in that seat, at that Speedway, on that day, and among those very people. I was being baptized into the religion of stock car racing in the storied waters of the Daytona International Speedway, a baptismal that some will say is the equivalent of NASCAR's River Jordan. I was astonished, excited, and amused, and oddly enough, even at that first race I felt the beginnings of a stirring of kinship with these people who surrounded me in various stages of sobriety and in all manner of attire and lack thereof. There were obviously blue-collar fans there, and there were some very well-to-do fans in attendance that day. Baltimore Ravens superstar Ray Lewis was there, and so was rapper 50 Cent. NASCAR, it seems, is the great equalizer.

Months of research had brought me to my very first NASCAR race. I drove the seven or so hours from Atlanta to Daytona, and I paid the fifty-five dollar (cheap seat) price of admission to the Grand Dame of NASCAR's soirees. I wasn't about to miss a thing. I was going to experience it all, throw myself wide open to it, and I'll tell you why: because getting to that first race became a challenge and a goal once I decided to uncover the mysteries of NASCAR for myself. As I mentioned earlier, I decided at that first race that I would attend several more that season. They're like potato chips. Once you get started, one is never enough.

The sum total of my exposure to racing before this adventure was a vague memory of an older cousin from Virginia stopping over in Atlanta on his way to a race in Talladega. I was probably eleven or twelve years old at the time. My mother told us that he was on his way to "the NASCAR," like it was a hush-hush disease or some other unpleasant condition. When I begged to go with him, she appeared shocked when she answered my plea with a stern "Absolutely not." I suppose that I grew up, then, with the vague idea that NASCAR was for "other" people, not for "us." It took me a few days to make that first drive to Daytona, but that was intentional too. I wanted to enjoy sightseeing along the way. On the Saturday before the race, I found a flea market in Daytona, where I discovered that you could buy anything from a tomato to knockoff designer sunglasses to a real hand grenade, which I'm assuming was not live, but I can't swear to it. At any rate, the market was one more adventure I added to my already growing list of "firsts." Then came race day Sunday.

While at my first race, I spent a fortune on a completely mismatched collection of NASCAR gear and paraphernalia in a pitiful attempt to look like I knew what I was doing. Instead of blending in with the crowd, I drew a lot of snickering, finger-pointing attention to myself. I even drew some rude remarks from other fans, and I couldn't for the life of me figure out why. Picture this: I had purchased and donned a #29 Kevin Harvick T-shirt, shedding my Kentucky Derby jacket for something more race appropriate. I selected a sun visor and sunglasses, two bobblehead driver dolls (caricaturing drivers #55 Mark Martin and #22 Joey Logano), and I ended up all decked out in a blinding collage of primary colors, the checkered flag theme and sponsor logos. Unfortunately, in the interest of Swiss-like fairness, I had assembled a hodgepodge collection of souvenirs and garb with almost every driver's name or number emblazoned somewhere on it. I looked more like Benedict Arnold's Christmas tree than a race fan after my first shopping spree that day, and that's why I drew "boos" from some fans as a result. Janice had to explain to me why I was becoming unpopular with just barely one race under my belt. You see, race fans are typically loyal to one, and only one, driver. Therefore, the visor must match the portable fan/water mister, which must also match the T-shirt, the bobblehead doll, the phone cover, and the souvenir beer mug. I clearly had no idea what I was doing. I remember Ed Clark's reaction when I told him a couple of months later what I had purchased, mixed, and matched in Daytona. He cringed—visibly—and then told me, "No, no. You can't do that. You need to pick a driver and stick with him." When I asked him how I might go about doing that, he smiled and said, "You can pick him for any reason you want. It can be for his looks, for his sponsors, for the car he drives, or for his personality. But pick one driver and stick with him. In your case, I'd pick a nice guy." That was good advice, and it was given to me by a man who's been in the NASCAR business for forty-one years, so I took it. For the record, after doing my homework I chose #15, Clint Bowyer, as my driver. He is in his mid-thirties, drives a Toyota Camry for the Michael Waltrip Racing Team, and is currently ninth in the Sprint Cup Series championship standings. He also has a bit of a quirky personality with a slight sarcastic edge, and he does some pretty cool charity work. With his big blue eyes, sandy blond hair, and scruffy beard, he's easy on the eyes too. I figure I have all my bases covered. Mr. Bowyer, make me proud.

Even though I had chosen Bowyer as my driver, I was unwilling to give up the #99 Carl Edwards visor I had bought (Edwards does a backflip out his driver's-side window every time he wins a race, then runs into the cheering crowd—how cool is that?). I stubbornly wore that visor, paired with my Kevin Harvick T-shirt. I fanned myself with my program and discreetly folded and tucked most of my other purchases into my race day bag. I had also bought a miniature checkered flag to wave and annoy my neighbors, and a red foam finger (which can be used for the same purpose, as it turns out, drawing the profanity I mentioned earlier). I also paid way too much for one of those big foamy beers in a fabulously tacky plastic take-it-home-with-you jug. I don't drink beer (at least, I didn't at the start of race season), but driver Brad Keselowski looks mighty cool in sunglasses, and therefore my beer mug with his face on it looked cool too.

The drivers and crew members in NASCAR are famously accessible to their fans, a trait that I found both puzzling and endearing. I walked around the racetrack, both inside (as close as I could get, anyway) and outside before the race, because there's so much activity going on that it's a challenge to catch and see everything. Sponsor products are being demonstrated, and giveaways abound. If I had known that all that swag gets distributed before the races, I would have brought my own shopping cart. Products ranging from mayonnaise to motor oil, soft drinks to potato chips are practically slung at passersby. My arms were full by the time I finally found my seat. On the way there, though, I saw a young mother ask a pit crew member, who was rushing to the infield, to sign her son's diaper. He obligingly agreed. *Sign his what?* I thought. *What will she do with the diaper later?* The possibilities really bothered me. I finally had to let it go.

There was a couple sitting a few rows behind me, alternately matching each other chug for chug and shot for shot with the encouragement of the crowd, until they both quietly and without warning passed out in their seats under the Florida sun—no souvenir visors, no sunglasses. There were some clouds in the sky that day, but still, I fought to suppress the neurotic, maternal urge to quietly sneak up on them and apply sunscreen to their relaxed and dangerously exposed faces and arms. I also prayed silently that neither of them would get sick before the race ended. I was still naively thinking at that stage of the game that my silly white suit might actually stay white.

Nearby, two security officers politely but firmly escorted a young man out of the Speedway for reasons unknown to me. Across the way, the "dot" who had gotten my attention earlier had bought herself another one of those souvenir-mug beers and was dancing and twirling as she inched precariously closer to the steep steps in the grandstands. I thought to myself as I continued to politely sip and occasionally gulp from my own now warming vat of beer, *Go home, Carole. You are much too neurotic to enjoy any of this. NASCAR is definitely not your thing.* Boy, was I wrong.

While I have just described a few fan antics in the reckless, hospitable, beer-drinking, entertaining crowd that surrounded me in Daytona Beach, I think this is also a good time to explain that there is another fan experience, one in which I vowed to indulge before the end of the season, and that's the suite experience. The suites, or "boxes" as they're also called, are plush, luxurious, roomy, air-conditioned glass enclosures that I glimpsed from a distance at Daytona. I made a mental note to wrangle myself a seat in one of those beautiful suites, by hook or by crook, before the season screeched to a smoking halt in the fall. I never did, but someday, I will. I could get used to a quieter, climate-controlled perch from which to watch all this activity.

I will reiterate that the Daytona 500 was a race I'll never forget, no matter how many more I attend in the future, and not just because of the fun, the sideshows, and the memorable lessons I found at every turn. First of all, the excitement swirling around driver Danica Patrick was positively electric. Patrick, who made history by becoming the first woman to win a Sprint Cup pole and to lead a lap in the Daytona 500, finished eighth in what's come to be called the Great American Race. She is a thirty-something brunette stunner with a modeling background who began racing in the NASCAR Nationwide Series in 2010. She has wowed crowds and girded attendance ever since with her exciting, competitive performances in what has traditionally been known as a man's game. Danica Patrick is blazing a trail, as are Johanna Long, Jennifer Jo Cobb, and other female racers; now the first female NASCAR pit crew member (she shadowed Clint Bowyer's pit crew in 2013 at Daytona), the beautiful and well-muscled Christmas Abbott, has strutted onto a stage populated mostly by he-men. Women who pioneer in murky and uncharted territory have my utmost respect.

In the interest of full disclosure, I have to admit that I had absolutely no understanding of the race I watched that day in Daytona. I saw loud cars driving in circles, and a lot of entertaining people. I knew that I had to learn about the maneuvering, the positioning, the passing, and all the other things that make a NASCAR race not just interesting, but exciting, to watch. The good news is that watching that February race made me want to learn more about the sport, so the research would be interesting to me. I wanted to understand what I saw on the track.

More than two dozen fans were injured that Saturday in February, the result of a jarring, horrific wreck that happened so fast it was uncanny. I saw many things I'll never forget as long as I live at that first race, but that one stands out for obvious reasons. Twenty-year-old driver Kyle Larson's (#32) Chevrolet Camaro went airborne during the final lap of the race, a result of those skilled, fast-paced driving maneuvers that sometimes garner drivers the lead, but sometimes result in a race-ending crash. Larson's engine and a tire flew through the air and reached the grandstand side of the catch fence in a wreck that ultimately involved not just his, but eleven other cars. I had never seen anything like that on such an enormous scale before. While Larson walked away and the emergency crew's response was fast and efficient, I asked myself a question that would again be asked of the powers-that-be of NASCAR: "Are fans safe at these super-events?" I understood after that day that aside from all the hoopla, press coverage, commercialism, and hype, racing is a perilous, no-holds-barred sport, from both the driver's side and sometimes, from the fan's side of the fence. When there is a wreck, there is the very real possibility for many fans that they may be caught up in it, getting injured or even killed. There is true potential for fiery tragedy and that, oddly, is part of the attraction to racing. Whether a driver loses control of his car and goes into the wall, or tires or engine parts go airborne and rain down on the crowd, the danger of injury is very real at a NASCAR race. Safety is at the top of NASCAR's list of priorities, but the possibility of getting hurt—or worse—will never be completely eradicated.

Every NASCAR race is an event, a super-happening with its own personality and characteristics, and when I went to my first race, even then I began to understand the fans' fascination with it all. I could see where the hype comes from and why it works in terms of dollars and attendance; in fact, I can't truly refer to all the buzz surrounding NASCAR as "hype" any longer, because the buzz, the enthusiasm, is genuine.

I will say this about attending my first live NASCAR race—as well prepared and informed as I thought I was, I fell woefully short of the "race day Carole" who would slowly evolve over the season. When I left Daytona International Speedway that day way back in February, I knew I obviously had to improve my race day strategy. Who goes to such an event by herself, and in that ridiculously inappropriate outfit? That wouldn't be happening again. No wonder people looked at me funny (though that wasn't the only reason, I'm aware). Next time, I would make sure my husband could go with me. I'd also invite some friends along, friends who know an awful lot about NASCAR and who had encouraged me from day one when I shared my idea for this exciting life adventure. Second, I knew without a doubt that I could master pulling off my own pre-race festivities—tailgating—the way I saw race fans do it in Daytona. Now I am very familiar with tailgating SEC college football-style. However, I had never seen anything quite like the tailgating that goes on at NASCAR. I decided that day in Daytona, as I walked past some partiers and was invited to join others amid hollers of "Show us your boobs!" and "You want to smoke some of this?" (I declined both offers), that I would equal or surpass some of the fabulously grand spreads I saw displayed by some very talented tailgaters. I did partake in some scrumptious food and some crazy cocktails while at that inaugural race, and I'll share those experiences with you later.

Third, while I thought I understood a good bit about protection from the Southern sun, I was wrong. I walked, or limped, out of the Speedway that afternoon with sunburned arms and aching feet. Long story short, those high heels may still be in the trash can outside the exit. I hope someone finds them, that they fit perfectly, and that that person thoroughly gets my money's worth out of them. I vowed I would never wear impractical clothing or shoes to another race. I was both mentally and physically exhausted at the end of that day, but I resisted the urge to throw away my stash of souvenirs, memorabilia, and licensed clothing simply because I didn't feel like carrying it all to my car, which after the race seemed to be parked somewhere in Alabama. I had the fleeting but panicky thought that I might never find it amid the campers, RVs, and trucks outside the Speedway, and I distinctly remember after the brief panic not caring all that much. I kept everything I bought that day, including my red foam finger, the tip of which had been torn off by a disgruntled fan. I was keeping my purchases, because each one already had significant meaning to me. I looked like a woman dragging herself off the battlefield with the spoils of war that Sunday afternoon, but I was walking away on my own power, a lot wiser, and most definitely intrigued. The most important thing I walked out with that day was the knowledge that I could very possibly enjoy this whole NASCAR thing and all the socializing and excitement that go with it, the stuff that I now call the race day "dance." I was already thinking ahead to my next race day trip, during that long, lonely barefoot walk back to my car.

Chapter Six

Every Race Has its Rules

There was one thing missing at that first race experience in Daytona, and that one thing became painfully obvious several laps into the race. As I confessed earlier, I had no idea what was going on down there on the track. When some fans cheered, others would shake their heads in disgust, and gasps of "Oh!" would ripple through the grandstands as if on cue every now and then. There's no getting around the fact that in order to completely soak in everything NASCAR, I had to understand the rules of stock car racing. There is precise method to the madness of hours of earsplitting noise, breathtaking near-misses, flags, lines, lightning-fast tire changes, and maintenance stops (also known as "pit stops"). All these things make up the precisely synchronized waltz that's hidden under all that grease and grime. NASCAR and the races are actually pretty easy to understand, once you get a basic grasp of the rules. Basically, and this is very basic, the guy who crosses the finish line first wins the race. Compared to football, I think stock car racing is a breeze to understand. I've been a football fan for more than thirty years, and I still don't get all the complicated intricacies of the game.

I have to admit that at the outset of my journey back at the beginning of the race season, I actually thought that NASCAR had no rules, with the possible exception of, "Stay inside the lines if you can." *How hard could it possibly be,* I thought, *to count five hundred or so laps, then just see which car crosses the finish line first? Better yet, just skip the race, tune into the eleven o'clock news, and find out who won that way?* Of course now I know that the uninformed "there are no rules" concept is not just silly, but even moreso, it's impractical, dangerous, and terribly naive.

The rules that NASCAR sets in place are established to maintain order on the track, to ensure driver, pit crew, and fan safety, and to support the points scoring system. Regulations regarding the cars are very specific to the *n*th degree and are designed to ensure safety and fairness, to the extent that that's possible. In fact, I learned while visiting the Hall of Fame that there are actual calipers, made by the various manufacturers, that fit over the entire race car to make sure that every measurement and distance on a car is legal. Those huge metal calipers reminded me of the body mass index calipers that physical trainers use on unsuspecting victims at the gym, the ones that pinch your arm fat to tell you just how out of shape you are. There are lines, boxes, flags, poles, pits, stripes, plates, and cups that have to be minded by drivers and their teams in every race. At the blistering speeds of a NASCAR competition, that's even harder than it sounds, because watching a race is much like watching toy cars spin around on high in a blender if you don't understand what you're looking at. Even as a more seasoned spectator, I still have a hard time keeping track of everything that's going on during a race. At any rate, the regulations do make sense of what otherwise looks like forty-three drivers circling a track at breakneck speeds with only the last lap counting for anything, which in hindsight was the sum total of my understanding at the Daytona 500. I'm embarrassed to say that I left that race not even knowing who had won. I didn't know that there had been twenty-eight lead changes or that the average speed was 159.25 mph. I had no idea that the pole speed (fastest qualifying speed) had been a scorching 196.434 mph. That probably explains why I spent the bulk of my time people-watching at that first race, because while the crowds at a NASCAR race are diverse and wildly entertaining, I had no idea that the race itself is also quite a show.

IN TERMS OF THE RULES, let's start with the basics. NASCAR races fall
into three tiers, somewhat like divisions in other sports. The Sprint
Cup Series races (formerly known as the Winston Cup Series)
comprise the top tier, the Nationwide Series (formerly known as
the Busch Grand National Series) makes up the second tier, and the
Camping World Truck Series (formerly known as the Craftsman
Truck Series) is the third tier. Think of the different series in terms
of baseball, in which there are AA, AAA, and major league teams.
Or think of Division I, II, and III college football. Those are loose
comparisons, but they convey the general idea. The series names
change when the series sponsor changes. Each series has its own
race dates, locations, and times. The best and simplest explanation I
heard for the reason NASCAR has the different series came from
some tailgating fans who were cooking a fabulous low country boil
(a spicy Southern/Cajun marriage of shrimp, crawfish, Andouille
sausage, potatoes, and corn) at Daytona. As we nibbled on shrimp
and cooled the heat of the Cajun spices with a chilled beer, the
couple explained the divisions this way: Drivers in the Sprint Cup
Series are thought to be the best of the best, and these races are
considered to be the toughest competitions in all of motorsports.
The Nationwide Series includes many drivers who are just starting their professional driving careers in NASCAR, and there are
some mechanical and structural differences in the cars that result in
overall slower paces during a race. There is also one fewer Nationwide Series race than in the Sprint Cup series. The Camping World
Truck series is actually pickup truck racing. NASCAR founded this
series in 1995, and while the original sponsor was Sears Holdings
Corporation (Craftsman), Camping World assumed that role when
Sears announced in 2007 that the company would no longer be a
sponsor. Therefore, the third-tier racing series has nothing at all to
do with camping as I had originally thought (I could not for the
life of me imagine how camping could be considered a competitive
sport, but as a journalist, I was open to the idea). Drivers can race in

one or all of these series if they wish, but they are eligible for the championship in only one of the series.

This may sound silly, but I didn't know that drivers are actually just one member of an entire racing team. I would never have guessed that NASCAR is a team sport but in fact, it is. Teams are made up of the driver, the car owner, the mechanics and engineers, and more—literally dozens of people. Think of the driver as the team's quarterback, or the player who gets the most attention. The driver typically gets the glory of a win and the blame of a loss. He is critically important, but there is a team supporting him. The owner is the boss, period. What he says goes. The crew chief brings experience and talent to the table. He knows how a car will perform on different tracks and under various conditions. He tells other team members how he wants them to perform their jobs at both the garage (the "shop") and the racetrack. He decides how everything about the car will be built and adjusted. He's the go-to guy. The car chief makes sure that what the crew chief wants done, gets done. The team manager represents the owner in the shop. Sponsors are also team members, paying anywhere between four million and ten million dollars for the privilege of sponsoring a team, and even more to sponsor a NASCAR cup series (like Sprint, Nationwide, or Camping World, currently). Team sponsors write the checks and market the team. There is a specialist (he takes care of the tires and everything to do with them), a pit crew, mechanics, spotters, fabricators, and engine builders. The guy who drives the eighteen-wheeler with its precious cargo of two race cars (the primary car and a backup) and their replacement parts—literally millions of dollars' worth of equipment—to every race is also a member of the team.

Something else that I could not wait to ask a person in-the-know was, "What in the world does it mean when race announcers say that a driver is 'on the pole'? That expression has always bothered me. Imagine my relief when I learned that what it actually means is that the driver with the starting pole position in a race (or "on the pole") is simply in the number one starting spot when the race begins. For one lap anyway, that driver is ahead of the pack. A driver earns that position by being the fastest during qualifying, which takes place in the few days or hours before the actual race. I was glad to get that cleared up, because it just didn't sound fair to me for anybody to have to sit on a pole for any length of time.

Every race is scored using a points system. In every NASCAR Sprint Cup series race, there are forty-three drivers, a fact that I found very interesting. There are not forty-four, and there are not forty-two. As with every other rule of the race, there is a reason for that exact number. Many years ago in NASCAR, there were some racetracks where there were no limits on the number of drivers competing. The egg-shaped Darlington Raceway in South Carolina is one such track, where there could be as many as seventy-five competitors in a single race. Imagine that; I pictured rush hour in Atlanta on the infamous, circular I-285, cars inching along at a snail's pace. Incidentally, Ed Clark shared an interesting piece of trivia with me about why the Darlington racetrack is egg-shaped. It's simply because there is a good fishing lake at one end, and the developer didn't want to drain the lake or change the track layout, so he simply made one end more narrow than the other. In Daytona, there could be as many as fifty drivers in a race years ago. Over the years and for various reasons, the number of drivers allowed in each race fluctuated. For a while, there were forty for the large tracks and thirty-two for the small tracks (not all racetracks are the same length). Sometime around the late 1980s to early 1990s, there were forty-two drivers competing on the longer tracks and thirty-six on the short tracks. Today, there are just forty-three, no matter the length of the track. It's a much simpler system. Every race on the schedule is worth the same number of NASCAR points, except the Sprint Unlimited in Daytona (until 2013 known as the Budweiser Shootout) and the Sprint All-Star race in Charlotte, which are not worth any points at all. Now here's where things get a little tricky. How can a race be fair if all the cars aren't lined up exactly even at the starting line? With forty-three cars in every race, that is of course impossible. Ideally, the fastest forty-three cars that show up for NASCAR qualifying during the prerace runs would start the race. But there is a system in place in order to help those teams that show up and perform consistently race after race, but may have had mechanical problems or even crashed during qualifying. In every Sprint Cup race except the Daytona 500 (which has its own set of qualifying rules), the top thirty-six spots are set by speed alone during qualifying races. NASCAR reserves a few positions for those drivers who may have had a problem during qualifying, so the next six positions (thirty-seven through forty-two) are set by points for teams that didn't make the race based on their qualifying time. Therefore, these teams are positioned at the start based on points, not speed.

Here's where things get even trickier. These rules leave one final spot open, the forty-third, which is known as the Champions Provisional. This final starting position is reserved for any former NASCAR champion who didn't qualify for the race any other way (either by points or by speed.) A driver can only use the past Champions Provisional once every six races. In other words, if a driver uses the Provisional option, he will have to attempt to qualify six more times before he can use it again. If there is no driver eligible for the Champions Provisional, then that spot goes to the eighth fastest driver that is not guaranteed a starting spot based on points. As convoluted as all this sounds, it seems to me that these regulations ensure two things. First, the best qualified drivers are able to compete in any given race and second, that a proven winner gets a shot at the Cup too. It's actually a very fair way to position competitors, in my opinion. Points are scored and races won based on a driver's position when the race ends. The scoring structure, implemented in 2012, awards points in one-point increments. For example, the most points a driver is awarded for winning any given race is 43, with the second-place driver winning 42, and so on. Drivers can be awarded bonus points for certain things, as well. For example, if a driver wins a race (43 points) and leads the most laps, he receives three bonus points for winning, one bonus point for leading a lap and one more bonus point for leading the most laps. With every race, the points are tallied, and drivers' standings can and likely will fluctuate after every competition. NASCAR made these changes to its scoring system in 2012 in the interest of simplicity and uniformity. Without drilling down into too much detail, NASCAR also has a playoff series called the Chase for the Sprint Cup. Twelve drivers will qualify to compete in ten postseason races, with the winner again determined based on points. This ten-race "tournament" follows a long and grueling season. Some drivers and crews like having the postseason competition, and some say that it's too stressful and deemphasizes an entire season's performance. In any case, NASCAR Chairman Brian France changed the way his father and grandfather structured the season by adding postseason racing in an attempt to make racing more interesting. News reports also stated that France believes playoff races give more drivers a shot at winning a championship. But again, one of the things I really like about NASCAR rules is that they can change in any given season, based on fan and driver input as well as safety concerns. For now though, the Chase is here to stay. Now let's say that it's race day, all the drivers are in position, and the announcer has uttered the traditional, "Drivers, start your engines" command, or something close to that. The green flag waves, and the rich gasoline-nourished rumble of the cars rolls over the crowd. As an interesting aside, former NASCAR driver and now race announcer Darrell Waltrip utters these famous words heard throughout the racing community at the beginning of every race he calls: "Boogity, boogity, boogity, let's go racin', boys!" Legend has it that when he was a driver, Waltrip was reportedly tired of hearing his spotter or pit crew chief shout, "Green! Green! Green!" at the beginning of every race in which he competed, so he switched things up a bit. At any rate, the command is issued and the cars roar and bellow, taking off in a managed, barely contained procession at first because of their starting order, weaving a bit, gaining speed until they are all flying around the track so fast that it becomes nearly impossible to keep up with them. The large numbers on the top and sides of each car help you keep track of your favorite driver. The shorter tracks make it even harder to keep up with what's going on, because obviously the laps seem to be faster. At least, the cars are jumbled up closer together, which makes things seem awfully fast.

The race is underway, and the skills of the drivers and their teams will be put to the test. They work together throughout the race, talking to each other, fine-tuning strategy and responses to other drivers' moves, always watching for trouble on the track, openings, and of course, the flags. I knew that a green flag waves at the start of a race and a checkered flag at the end, but there are a lot of flags that may be used in between those two. It's the team's spotter who is responsible for keeping up with the flags and communicating their message to the driver. During the heat of a race, the spotter's words are critical to the driver, with those words really being the only way a driver knows what's going on outside the noisy interior of his car. At around 180 mph, there is no room for guesswork. Remember too that the driver really only has a clear view of what's directly in front of him. His helmet, the safety bars around his head and neck, and the banking of the tracks limit his visibility. His spotter, positioned in an area high above the grandstands, has a view of the entire track. A driver's spotter literally talks him through passing other drivers, warns of upcoming trouble such as debris on the track or a wreck, and the waving of the different flags from the flag stand. Everyone knows that the green flag means "go" and the black and white checkered flag means "race over." But what about the yellow, or caution, flag? When it's waved, a pace car enters the track and leads the drivers at a slower predetermined pace, depending on the track and the race. The hazard could be that the track is completely or partially blocked, that there is debris on the track, that it's started to rain, or even that an animal has wandered out into the fray. In 2009, a gray fox drew a yellow flag when he somehow wandered onto the track during a Nationwide qualifying race. There have been caution flags thrown because of deer, rabbits, ducks, runaway tires, and in one case, the yellow caution light hanging overhead actually came loose and fell onto the track. The caution flag was thrown because the caution light had fallen. In the case of the fox, he outmaneuvered the staff, who were trying to corral him by ducking into what appeared to be a secret trap door alongside the lip of the concrete track. The "critter," as the announcer referred to the fox, was a little rattled,

When rain makes track conditions too dangerous, track dryers are brought out to dry the surface.

but unharmed. That day was indeed his lucky day. At the April 2013 Richmond night race, the grounds crew at Richmond International Raceway forgot to turn off the sprinkler system before the green flag waved. On lap 160 of 400 of this fast-paced short-track night race, the sprinklers popped up and soaked the track, drivers, fans, and crew members. The yellow flag waved furiously, and the forty-three cars then ran at caution speed until the track was dry. It really is true that on any lap in any race, anything can happen. When the pace car enters the track during a caution, drivers are locked into position, no passing allowed. Most drivers take advantage of a caution, ducking into their pit for a tire change, to fuel up and to get adjustments made to their car. When the caution is over and the problem cleared, the green flag is again waved, and it's every driver for himself. There is a red flag, a black one (that one means that a driver is in trouble, typically for a rule violation), a white one (this one warns that there's one more lap until the race is over and if you're watching from home, your nap is almost over), and a few others, but my hands-down favorite is the blue and yellow flag. In essence, when that one is waved at a driver, it means that he should move over and let faster cars pass. Depending on the race, the driver doesn't even have to heed the warning. Doing so, however, is considered proper race etiquette. For some reason, I find this to be very funny. These drivers are whizzing around a track at breakneck speeds to win themselves anywhere from several hundred thousand dollars to more than a million (unlike football and baseball, NASCAR keeps its drivers' salaries a closely guarded secret), and politeness dictates that one should move over to let someone else get ahead. I'll bet that a Southerner came up with that one—manners first, always. You may be asking yourself right about now *How on earth can a driver traveling at nearly 200 mph see a flag, much less know that the flag in question is meant for him and not the guy that's right on his bumper?* Good question. There are flag men, spotters, and of course the pit crew who keep their eyes trained on critical race components and communicate them to their drivers. When fans listen in on their scanner headphones, these are the conversations to which they're listening, because the drivers and team members are wearing them too. These team players are crucial to the drivers for several reasons. As I mentioned earlier, the pace on the racetrack when drivers are involved in full-on competition is faster than the human eye can track, or at least faster than my human eyes can track. Imagine driving at those speeds and having to take your eyes off the action for even a split second every time you pass the starter stand to see if there's a flag waving, and what color it is. When a driver on a racetrack looks to his right, he sees asphalt and perhaps a wall, if he's close to the wall. To his left, he sees a multicolored blur of cars, garages, people, and grass. The tracks are banked high on the outside, and the angle is much steeper up close than it looks on television. Without spotters and other team members communicating to the drivers, they would practically be driving blind. To me, it was a confining and even panicky feeling when I had my own personal NASCAR driving experience. Drivers need to be told many times when it's safe to make a lateral move or to pass another driver, because their perspective on the overall field is wide, not narrow.

NASCAR is open-minded about its rules and their purposes from season to season. NASCAR officials have proven time and

again that they will listen to both drivers and fans if they feel that
a rule has outlived its purpose or can be made better by changing
it. The racing rules of competition, then, are fluid and shifting, but
have safety and the good of the sport at the center of them all.
For instance, since 1988 at both the Talladega and Daytona racetracks, a device called a restrictor plate must be installed on every
engine of the cars competing in a race. This shiny device with four
holes drilled into it (I actually held one in my hands at one of the
super-garages in Charlotte—they're great for doing a quick touchup
of lipstick and hair) is placed between the carburetor and the intake
manifold of the engine to reduce the flow of air and fuel into
the engine's combustion chamber. Simply put for all of you out
there who know about as much as I do about a car engine: restrictor
plates reduce both horsepower and speed. They were implemented
and their use made mandatory in 1988, following driver Bobby
Allison's 1987 crash into a retaining fence at Talladega Superspeedway. According to the accounts I have read about the incident, an
old newspaper clipping I found dated May 4 of that same year,
the right rear tire of Allison's #22 Buick exploded, causing the
car to slide sideways, go airborne, and tear a 150 foot hole in the
grandstand catch fence. The accident injured several spectators and
endangered hundreds more. It was this incident that got the attention of the powers-that-be in NASCAR, as Allison had reached
210 mph in several earlier laps of that race. Bill Elliott ("Awesome
Bill from Dawsonville") set a record in his #9 Ford by running
a lap in that same race at a blistering 212.809 mph. That May 3,
1987, race, the Winston 500, was the last unrestricted race ever held
at Talladega. Restrictor plates are now required for cars racing at
the longer tracks of Talladega and Daytona following the horrific
spectacle of Bobby Allison's crash that day. Some experts contend
that, with the speed that today's cars can reach, drivers would easily
exceed 225 mph on these super-tracks. Many fans I've talked to say,
"Let'em race! That's what it's all about!" NASCAR, however, must
consider both driver and fan safety to take priority over wide-open
competition with no limits whatsoever.

There is another faction out there that will argue that restrictor plates are actually the cause of multi-car accidents. Since the devices reduce speed by about 10 mph, the result tends to leave the field of forty-three cars (remember, always forty-three) bunched together too tightly. If one of those cars crashes at even the reduced speed of around 190 mph, it will likely cause several other cars to crash along with it. While the jury may remain out on the restrictor plate regulation, the rule remains in place at Talladega, Daytona, and most recently, at New Hampshire Speedway. There really is order to the chaos, once the rules are understood.

Yet another term that I learned early on is bump-drafting, which refers to a fascinating combination of two NASCAR racing maneuvers that is every bit as controversial as the "dunking versus sprinkling" argument is to Southern Baptists and Methodists. They're fighting words. The technique is exactly what it sounds like—a driver on one of the long superspeedways can use his car to bump another car on the rear bumper, which can lead to both cars drafting together and gaining speed. Of course, it can also result in the bump-drafting car knocking the bump-ee out of the way and possibly into the path of other cars jockeying and maneuvering at smoking speeds. It all depends on the angle of the nudge. Obviously, drivers have very passionate opinions about the use of this technique. When it's used too aggressively, bump-drafting can be extremely dangerous.

In 2006, after the Daytona 500 was run, driver Tony Stewart (#14) voiced his concerns to NASCAR officials about the use of bump-drafting being out of control and dangerous. Stewart even warned of the possible death of a driver if the practice continued unchecked. NASCAR's reaction that following fall was to warn drivers during a prerace meeting that if they bump-drafted, NASCAR would park them and the car they rode in on—no excuses, no questions asked. The unintended result was a messy, dull race that still involved wrecks that occasionally sent cars flying, sometimes upside down. Obviously, neither the fans nor the drivers want a boring race, so the ruling was reversed and drivers were essentially admonished to use good judgment and proper etiquette in the use of bump-drafting. In other words, "play nice, boys and girls." Currently, the new Generation 6 cars (Gen 6 is the newest generation of NASCAR automobiles, constructed more true to their manufacturers' stock) have bumpers that aren't all that well suited for bump-drafting. Surprisingly though, the practice is not forbidden. In January 2013, during some of the test runs at Daytona International Speedway that lead up to the Daytona 500, the first real shot a driver took at bump-drafting was disastrous. Dale Earnhardt Jr. (#88) nudged the rear bumper of driver Marcos Ambrose's (#9) car, a move that started Ambrose sliding and quickly dominoed into a twelve-car smashup that ended the test session for several driving teams. While no one was hurt in all that mess, the cars took quite a beating, and that, my friends, made bump-drafting that day an expensive little maneuver for several teams. As it stands right now, today, bump-drafting is allowed except in turns at Talladega and Daytona. Of course tomorrow, that could all change. NASCAR will try something, weigh the results with respect to the safety vs. entertainment factor, and then completely reverse a policy if it sees fit. As a woman, I can fully appreciate that approach.

Chapter Seven
The Day I Got Hooked

The Daytona 500 whetted my appetite for speed. It left me wanting another taste of stock car racing, a fact that still surprises me. I am, or so I thought I am, allergic to adrenaline. I was never a thrill junkie. My idea of living dangerously is letting myself run perilously low on my favorite hairspray or paying full price for a pair of shoes when I feel sure they'll go on sale in a week or two. I do not like speed or danger, extreme heat, or noisy crowds. Having said that, I just described what a NASCAR race is, in a nutshell: speed, danger, extreme heat, and huge crowds. However, as a journalist I learned a long time ago that going right to the source for any story is the only way to get the facts straight and truly relate an experience to the reader. I'm also a hands-on learner, so I knew without a doubt that I was going to have to go about this the old-fashioned way: by doing things myself. I could not hire someone to attend races for me. There would be no interviewing someone else who had driven a race car before and would be willing to describe the experience to me. As foreign and at times terrifying as these affairs would prove to be, I decided that I would have to be the one to experience them.

No matter how I approached my NASCAR education, I kept coming around to one inevitable, worrisome conclusion. I knew that I would eventually have to climb inside a race car and experience the speed that I had been reading and hearing about, the speed that I had seen for myself on television races, the speed that had shaken the grandstands and caused a stupendous crash at Daytona. Being inside the race car while it flew around the track, I just knew, had to be a different experience than simply watching the cars race, hearing the growl of the powerful engines, smelling the exhaust. Reading about g-forces, elevated heart rates, and reckless thrills will never be the same as living and feeling them, and I knew that I'd have to dive in headfirst in order to really share what it's like to be in a car that's traveling about 175 mph. This inevitable conclusion was not one that I relished; the more I thought about actually doing it, the more my worry and fear gnawed at me, threatening to mushroom into full-blown panic. Truth be told, I investigated the possibility of using a driving simulator instead of driving a race car myself. However, the purist in me eventually won out. I closed my eyes, held my breath, and booked myself a six-lap NASCAR Driving Experience at Atlanta Motor Speedway. Once I took care of the business of booking the Experience, I shoved the looming prospect all the way to the back of my mind until the driving date rolled around; otherwise, I don't think I could have actually gone through with it.

My husband, my son, and I arrived at Atlanta Motor Speedway bright and early on the morning of my NASCAR Driving Experience, the butterflies having emerged from their little cocoons in my stomach and already racing laps in there. I don't think I slept a wink the night before, obsessing about whether we had updated our will since my husband and I had gotten married. The drive to the speedway had taken about two hours, and I spent that entire time quiet and in deep thought, imagining and playing out worst-case scenarios in my creative mind. I had staged the standard "driver crashes into wall and bursts into flames" scene, which ended with the emergency medical team pulling me out of the car, looking like a piece of crispy overdone bacon, my husband and son wailing and mourning the senseless loss. I imagined the high-waisted mom jeans I was wearing splitting from front to back as I awkwardly climbed through the car window, shamelessly displaying my mom underwear for the pit crew and drivers to see. Remember, NASCAR stock cars do not have doors that open and close. Access is through the windows only.

The day before the scheduled date of my drive had been stormy, with tornadoes twirling and spinning throughout the southeast, chewing up unsuspecting little towns in their paths. That meant that the track in Hampton would be wet, so of course I took that thought and ran with it. I did not know at the time that NASCAR was using a very expensive system of trucks and equipment that essentially blow-dries, vacuums, squeegees, and maybe even flat-irons a track bone-dry after a rain. The Air Titan™ system, developed at Brian France's direction by NASCAR Research and Development, reduces track drying time by as much as 80 percent. It was introduced in 2013. No, I did not know all that, so in my mind, a wet track meant sure death for me, as I imagined myself spinning around and around on the track until I reached the same end as before—smashing into the wall and bursting into flames. Every one of the nightmare scenarios I conjured up ended the same way, with me hitting the wall at some point and catching on fire shortly thereafter.

The two-hour drive that beautiful spring morning had given the worrywart drama queen in me plenty of time to get fully into character. My husband and son had to have been wondering why I was not as talkative as I usually am, but if they were, neither of them said anything. Perhaps they were enjoying the abnormal quiet, but I suspect they understood full well that I was terrified. I had worked myself up into quite a state, eventually having to go to that place in my head in which it almost seems that I'm a spectator, dispassionately watching the events in my life unfold from a distance. I would have backed out of the drive completely, forfeited the entire cost of a little over two hundred dollars gladly, if my husband and son hadn't been with me that day. They would have been so disappointed had I not followed through; somehow, on some level, I felt that my honor and that of fifty-something suburban moms everywhere was on the line.

The approach to Atlanta Motor Speedway is an impressive one, with a long flag-lined road opening onto a panoramic view of the home of one of the fastest tracks in NASCAR. The sprawling facility, which sits on 887 acres just twenty-five minutes south of Atlanta in Hampton, Georgia, has not always been so extraordinary. In fact, the facility struggled financially in the 1960s and 1970s, but in 1990, Bruton Smith bought it. A few years and tens of millions of dollars later, sweeping changes were made to it, with the then-thirty-six-year-old track being almost completely rebuilt. Ed Clark told me about many of those changes: "The frontstretch and backstretch were swapped, and the configuration of the track was changed from oval to quad-oval. That project made the track one of the fastest on the NASCAR circuit." The entire front suite level was rebuilt, also. Sadly, a devastating F-2 tornado tore through the speedway in 2005, doing an estimated forty million dollars in damage. A herculean day-and-night effort began immediately to make repairs and get the track ready for an October race—the Bass Pro Shops MBNA 500—that had already been scheduled. Round-the-clock workers whipped the track into a condition that was even better than before, and with three weeks to spare. The recently built, tastefully elegant "1500 Tara Place" condominiums overlooking Turn Four are also where the corporate offices are located. Atlanta Motor Speedway really is a beautiful facility.

Anyway, my Driving Experience packet included instructions that directed participants to drive through the infield tunnel at the track and park right out beside pit road. The security guard at the gate, a friendly older gentleman, checked us in and joked with my husband about how terrifying the drive would be for me. He asked me whether I had brought a change of underwear and some smelling salts with me, just in case. The guys laughed and joked, exchanging good-natured remarks at my expense, and all the while I played out one devastating scene after another on my mental movie screen, each one ending in my demise by fire, impact, heart failure, or a combination of all three. Of course, once we parted ways with the guard and topped the first hill on the way to the parking area, the first two vehicles I saw were a fire truck and an ambulance. My eyes widened. *Why are they here? Am I really going to go through with this?*

The NASCAR Driving Experience, just as the other driving experiences offered to fans at various NASCAR tracks, is one in which participants can drive an actual race car or ride along with a professional driver. While my original intent was to drive the car myself, a recent knee surgery necessitated that I ride, not drive, that day as a safety precaution. Fine by me. I quickly calculated the lessened odds of my burning alive in a fiery crash, now that a professional driver would be in charge. Better, but not great.

We had seriously overestimated Atlanta traffic that Friday and had arrived at the track about two hours earlier than we had planned. I checked in and presented my six-lap driving pass, signing the voluminous waiver that basically stated that I was committing this ridiculous act and tempting fate of my own free will, and anything that happened to me was ultimately my fault and responsibility, and shame on me. Prepared to wait a couple of hours for my number to be called, I was surprised when the friendly woman with a megawatt smile took my pass said, "Well, you're early, but if you'd like to go now, step right over there," pointing at the track with six or eight race cars lined up in a menacing row, just itching to come alive, stretch, and growl. *Go now?* I thought. *You mean, now?* The cars had names of well-known drivers displayed on them—names like "Clint Bowyer" (#15), "Darrell Waltrip" (#17), and "Jimmie Johnson" (#48). They all just crouched there in a line, daring me to step closer and "go now."

The woman gave me an official-looking lanyard and guided me toward the waiting cars, pit crew members milling around them and tinkering with this and that. I was in a rubbery fog, my mind protesting from far away that no, in fact, I would not like to go over there now. I would like to go home. I would prefer to read about this experience, thank you very much. But my feet were moving, and my husband and son were watching, and then I was standing there under the tent pulling on a sanitary paper hair sock, then getting fitted for a helmet. *My hair,* I thought. *Why did I bother to do my hair this morning? It will be smashed by the helmet, then burned completely off when I crash into the wall.* I believe, at that point, I may have actually been delirious.

Some of the other Driving Experience cars lined up, waiting for drivers.

Garrett "The Gambler" Edmiston was the pit crew chief that day, and he couldn't have been friendlier. Maybe he sensed that I was nervous (if fear has a smell, I must have reeked of it) as he joked and chatted, doing his best to put me at ease while fitting my

Ed Clark demonstrating proper use of a helmet and HANS (Head and Neck Support) device.

helmet. My eyes darted to the right as Garrett tugged and adjusted the straps, and I took note that the professional driver patiently waiting for me in the car was wearing a fire suit. All I would be sporting during the ride was a helmet, a T-shirt, nicely done hair and makeup, and my mom jeans. *Please, God. Please let the stitching in the seat of my jeans hold when I climb in that window.* Oh, how we lower our sights when faced with a crisis—from worrying about a fire suit to praying that the crotch of my jeans had been double-stitched by a conscientious factory worker somewhere overseas. Why did I wear jeans anyway? What was I thinking? That faraway spectator's voice protested indignantly alongside that hasty prayer, *Wait a minute! That can't be right! Where's your fire suit, Carole? Say something!* I've even been told that the drivers wear fire-retardant underwear, but I was too embarrassed to verify that with Garrett. I did remember, with crystal-clear clarity, the factual account I had read somewhere along the way about fire suits and their importance to drivers (and passengers, I presume). In the late 1950s, maybe early 1960s, drivers were required to dip their race clothing in a solution that would help fireproof the fabric. Drivers could still wear whatever clothing they chose, as long as the clothes were treated. Driver Edward Glenn "Fireball" Roberts (nicknamed "Fireball" because of a burning fastball he threw as a pitcher for the Zellwood Mudhens) was allergic to the fireproofing solution. Since the chemicals aggravated his asthma, he brought a waiver from his doctor to the World 600 Race in Charlotte that day and was exempted from having to treat his clothes with the solution. On Lap Seven, drivers Ned Jarrett and Junior Johnson crashed and spun out, and Fireball crashed trying to avoid them. His Ford slammed backward into the inside retaining wall, flipped over, and burst into flames. Some accounts report that witnesses heard Roberts screaming, "Ned, help me!" from inside his burning car. Even though Jarrett rushed to try to pull Roberts out of the blazing inferno, the doomed man suffered second- and third-degree burns over 80 percent of his body and was airlifted to a nearby hospital in critical condition. While he stubbornly clung to life for a few precious weeks, Fireball finally succumbed to pneumonia and sepsis, and he died on July 2, 1964. Ironically,

Roberts was one of the first drivers ever to wear a standardized uniform that day. It looked like coveralls but unfortunately, it was not flame retardant.

My mom jeans are not flame retardant, either. Apparently, the thinking is that you won't burn up in a crash if you're in the passenger seat. What other explanation could there be for my not having a fire suit? Still, my feet moved obediently toward the car at Garrett's instruction, in spite of all this praying and protesting. Now ever since I booked my NASCAR Driving Experience, I worried about a lot of things, but one real question I kept coming back to was whether my knee would bend enough to fit in the car window. The surgery I had four months earlier had left it stiff and swollen, despite hours of rigorous physical therapy. As I approached the blue car, marked with the Number 48 and driver Jimmie Johnson's signature above the driver's side door, I said another quick prayer. *OK Forget about the hair and the fire. Please, dear Lord, let my knee bend enough to get it in that window. Please don't let it hurt too much, and please don't let me scream, fall out backward, or otherwise embarrass myself.* A tall order, I know, but I believe in a big God. I had nervously jabbered to Garrett

Mechanics fine-tuning the #48 Jimmie Johnson Lowe's car in which I rode.

the fact that I had had a recent knee surgery, and in a blink, he had two strong young men from the pit there to help me climb in the passenger-side window. They instructed me to put first my left leg in the window, then sit on the window ledge, then fold my right leg in. I did as they instructed and for a second, I thought, *It's not going to bend enough. I'll have to back out. How humiliating!* Then, in one swift and slightly jerky movement, my right knee gave way just enough for me to slip in the window, and I slithered down into the seat with a graceless plop and a deep sigh of relief. As the young man secured my neck roll and buckled me securely into the seat, I looked over at Jay Foley, the lucky guy who drew the short straw to be my driver for the day. He had big blue eyes and eyelashes that went on forever. I didn't have time for any of that though. This was serious business, and I needed critical information. "Hi, I'm Carole. What's your name?" I asked, my only thought being, *I need to know whose name to scream as we hit the wall and burst into flames.* "Hey there. I'm Jay," he answered, popping the steering wheel into place on the column (it's removable, and that made me more nervous for some reason). Just as I opened my mouth to ask him where my fire suit was and how long he'd been driving these cars, Jay started the engine. I don't know whether it's medically possible for your heart and your stomach to switch places, but that's sure what it felt like as the 600+ horsepower engine in the blue Chevy Impala SS roared to life. A typical race-ready NASCAR engine reaches more than 9,000 rpm during a race; the car I was sitting in reaches about 6,000 rpm, according to Garrett. Compare that to a standard automobile like my safe crossover SUV that reaches 2,000 to 3,000 rpm. That should offer some idea of the intensity, the sound, and the power of the engine. Allegedly, this car had been sissified for people like me. I glanced over at Jay for reassurance, I suppose, but his head and neck were securely in place between the steel bars*(Hey, where are my steel bars?).* The shield on his helmet was jet black, his face unreadable. No reassurance there. I took in my surroundings, eyes wide and breath quickening, as any possibility of making small talk in an attempt to delay the inevitable ended with the starting of the engine. I looked around for something on which to get a firm death-grip and gritted my teeth. The inside of one of those cars is surprisingly bare, at least to the untrained eye. I'm not sure what I expected to see, but I was intrigued by the fact that there was no leather, no carpet, no upholstery to speak of, save for the tough black covers on the seats. In fact, the floorboard and doors are bare metal, and the stick shift is just that, a long, black stick. Bare, gray steel is visible here and there. A removable black rubber net covers the windows on both sides for driver security. *Why is it, again, that I don't have a fire suit? Did anybody ever answer that question?* Jay revved the engine a couple of times, a loud, somewhat intimidating and, yes, kind of exciting sound. He shifted into first gear and we took off, leaving pit road and to my surprise, lazily weaving back and forth a few times. I thought he might be doing this to scare me a little. *This isn't scary at all,* I thought defiantly, but in fact Jay explained later that there's a reason for weaving like that. Doing so warms up the tires and gives him a feel for the road. I was his first crybaby of the day, remember. Several drivers have since told me that no two tracks feel the same, and no single track ever feels the same from one day to the next. Once he was through weaving though, and got onto an open stretch of track, he began to accelerate, deftly

Jay Foley and me before my Driving Experience, safety net up on window.

shifting through the gears as an opera singer might shift up and down through the musical scales. As we went into the first turn, I prayed a fervent and heartfelt prayer that, as I recall, went something like this: *Please God please God please God. . . .* Approaching a high-banked turn at about 165 mph is both terrifying and exhilarating. It looked to me as though Jay was heading straight into the wall, intent on playing out my first "worst-case scenario," but at the last minute he gently nudged the wheel. The turn created such force that I felt myself leaning hard to the left, on a wild rollercoaster, only without the assurance that we would not hit the wall and burst into flames. In the turn, it felt as though the car floated, like we were drifting, instead of gripping the asphalt. To my right, the white wall whizzed by, seemingly close enough to reach out and touch. Atlanta Motor Speedway is a one-and-a-half mile track (1.54 miles, to be exact). At 165 mph, which is the speed at which we topped out that day, it doesn't take very long to drive six laps. Before I knew it, my ride-along experience was over, with Jay gearing down and slowly easing the car back onto pit road. Once he shut the engine off, a crew member reached into the window, holding onto the door, and easily pulled the car into position to await the next passenger. The cars are built surprisingly lightweight—the new Gen 6 cars are even one hundred pounds lighter than the Car of Tomorrow—in the interest of maximizing speed.

I sat still in the seat for a minute, wiping the tears that had streamed and blown from my eyes straight back toward my hairline. I wasn't crying, though. Were they tears of exhilaration? I think so. All my limbs were intact, though I did have to pry my fingers off the edge of the seat, and all of my clothing was dry. *Thank you, God, for small favors.* Another prayer answered. Of all the things I expected to feel when the six laps were finished that day, the last was disappointment, which is exactly what I felt. *It's already over?* I thought. *But I was just starting to have fun.* I climbed out of the car, kissed the track and took off my helmet, grinning from ear to ear and making sure my boys had both been watching. Garrett met me halfway between the track and the stands to let me know that he had secured permission for us to take some photos of the cars inside and out, and to take a look under the hood. Even for someone like me, whose first thought was that the engine looked like shiny chrome intestines, what crouches under the hood of a NASCAR race car is an impressive sight to see. The engine is huge, and it's very shiny.

Before the entire experience was over that day I knew, without question, that I was hooked. For me, the Queen of Careful, to climb into a race car, fly around a track at 165 mph, and enjoy it—well, that was just crazy. I didn't think I had it in me.

I had a wonderful experience that day in Hampton, Georgia. From the funny, wise-cracking security guard, to Garrett, to Jay Foley and Bill Zacharias (the other professional driver at the track that day), all the way up to Mr. Ed Clark himself, every single person I met went out of his way to accommodate me and make me feel comfortable. Bill has been a lead driver with the NASCAR Driving Experience for seven years. Jay is a professional driver who travels the country competing in races. He lives in Louisville, Kentucky, and his love of racing sprouted and grew just like the love that so many other drivers has. In Jay's case, he grew up around his dad's salvage yard, his dad himself a racer at the Fairgrounds Motor Speedway in

Jay Foley and me after my driving experience.

Kentucky. "I bought my first car when I was fourteen, a sixty-seven Plymouth Valiant. I paid twenty-five dollars for it. For three months, I drove my dad and all the guys at the salvage yard crazy, racing around in that thing like it was a go-kart. It was a great way to grow up," Jay said of his childhood and love of speed and cars. These days, he travels about three weeks out of every month, getting "seat time" at different tracks, driving just about anything that someone wants him to drive. I think of him as a freelance driver, if that makes it easier to understand. He was in Atlanta that week getting "seat time," driving people like me around the track and racking up more experience. When I asked him the question that I eventually asked every driver I had a chance to talk with, he gave me just about the best answer I heard, or at least, it was probably the most honest answer, as to why on earth he loves to race. "I remember my dad putting me in a car in a field, when I was twelve years old. He just cut me loose, for hours. That's when that feeling started—that feeling that you're in total control, a part of the car. I know when it's too loose, too tight, when there's something wrong with one of the tires; I can just feel it. I can close my eyes and still remember that day," he reminisced, grinning all the while. His answer was simple: he loves speed, and he loves racing. He got it honest, that's for sure. I had to laugh when he told me the following story about his dad's love of racing. "In nineteen sixty-two, my dad was fighting for the championship at Fairgrounds Motor Speedway. At the time, my mother was pregnant with my older sister. Well, it was race day, and my sister decided she was coming out. My dad took Mom to the hospital, my sister was born quickly, and my dad said, 'Honey, I have to go to the racetrack.' Later that night, my dad's team won the championship." I wish I had known all that about him before Jay Foley took me for a spin around that track in Atlanta; I might have felt less panicked. Doubtful, but I might have. My husband, my son, and I hung around the track for a couple of hours after my ride, watching other participants who had paid for the experience of driving a stock car themselves around the track (they were all wearing fire suits, by the way). I talked to a few of the waiting drivers, each one as excited as the next to have this experience of a lifetime. Some had been given gift certificates for the drive for their birthdays; others had saved and paid for the experience themselves. One gentleman, who was eighty-two years old, had arrived earlier that day with his eighty-year-old wife. While he took the driving class with about forty other soon-to-be-drivers, she went right along with him. While he was standing in line awaiting his turn to drive, she snapped photo after photo. When it was finally his turn to drive, she sat in the stands, smiled sweetly, and continued to take pictures. She and I talked for a while before her husband got behind the wheel, and she explained to me that her partner of fifty-nine years was a two-time cancer survivor. He had always wanted to drive one of those powerful cars at a NASCAR track, open it wide up, and when he got the news a few weeks earlier that his cancer had been declared in remission, he made up his mind to do it. The elderly couple had driven in from Knoxville, Tennessee, making a long weekend of the unforgettable experience.

When we left Atlanta Motor Speedway that day, I felt like a changed woman. Oh, I know that may sound cliché, maybe even a little corny, but it's the truth. Before embarking on my quest to understand the cultural phenomenon of NASCAR, I would have never, in a million years, voluntarily done what I did that day. Never. I was too cautious, could cook up too many "what-ifs," to ever let go enough to do something as risky as I had just done. When we drove out of the infield tunnel and headed back home, I had a grin on my face that I couldn't suppress, even though it made me look a little crazy. I had done it. Now I could honestly say that I understood exactly why speed is such a thrilling rush, why the sound and feel of a car like that are so exciting. It's the element of danger. It's the closest most of us will ever get to flying and feeling it, to experiencing the force and sheer breathlessness that speed imposes on our minds and bodies. I liked it a lot.

We drove through the heart of downtown Atlanta on our way back home to the suburbs, and for those of you who are familiar with Atlanta and her delicacies, I'll share this too. We stopped at the Varsity for lunch while in the city. Before that day, I hadn't eaten there in probably fifteen years or so, ever since I responsibly decided to safeguard my arteries from tricky villains such as saturated fats, sodium, and the resulting plaque buildup. Answering the famous question that Varsity carhops have been shouting at patrons since 1928—"What'll ya have, what'll ya have?" I ordered two chili dogs and onion rings, and I ate every last bite. What a day.

Chapter Eight

The Evolution of the Sport—the Players

While I was devouring my chili dogs and relishing every bite, grease and all, a thought occurred to me. Have the cars in NASCAR racing always run that fast? Has the ride always felt like that, gliding in the turns and gobbling up the asphalt in between them? As impressive as the cars of today's NASCAR are, I don't think their strength and engineering can truly be appreciated until the big picture of the history and evolution of the sport is understood. For many years, NASCAR was considered the "poor man's" version of racing, while Formula One racing carried with it an air of sophistication and wealth. Mention "NASCAR" to many people, and they associate both fans and drivers with being a bunch of rednecks racing plain, everyday stock cars around in circles. It has long been considered "everyman's" sport for that very reason. While the mention of Formula One racing conjures images of sleek, futuristic-looking race cars, NASCAR's stock cars look an awful lot like any car sitting next to you at the traffic light on the way to work. There's nothing sleek or slim about them. In fact, one broadcaster likened the challenge of making a stock car go faster to "trying to sling a brick through some very resistant air."

Make no mistake, though. NASCAR's hayseed, country bumpkin image has been left far back in the dust, as today's cars are engineered and built with both big money and cutting-edge technology. The drivers are no longer former moonshiners or guys with a Sunday afternoon death wish. They are well-trained, physically fit men and women, just as scientifically suited for this job as the cars that they drive. NASCAR has thus far been able to maintain a balance between the contradictory worlds of science and good ol' boy camaraderie; while you'll find drivers, mechanics, announcers, and media milling around the infield before a race to do a very specific job, you'll also hear, "Hey, how y'all doing?" and "How's the family?" The hayseed reputation is fading, but the family atmosphere, some say, never will. I had the honor of interviewing Rex White, a 2013 Hall of Fame nominee, another North Carolinian, and an eighty-three-year-old NASCAR legend who raced from 1956 until 1964 and won the Grand National Championship in 1960. White told me that his first competition stock car was an old 1937 Ford for which he and his wife saved and paid six hundred dollars, but he won the 1960 championship in what became his trademark gold and white Chevy. Born during the Great Depression and suffering polio as a young boy, White is quick to say that as a young man, he saw cars strictly as transportation, not as symbols of an upcoming billion-dollar sport. He and his mechanics didn't do a whole lot to that first old Ford, either, to get it race-ready. The biggest purse White ever won in his career was thirteen thousand dollars in 1960. As a contrast, I'm not sure I've heard of a disciplinary fine being handed down by NASCAR these days for less than twenty-five thousand dollars. I had read a quote somewhere before talking with the racing legend, in which White said one of the biggest lessons he learned in life was how to conquer fear. I asked him what he meant by that, and he said very plainly, "There is no room for fear inside a race car." That's a good thing, as there wasn't much between White and that hardscrabble track back in 1960 other than a car that rolled straight off the manufacturer's line with some mechanical adjustments here and there; fear would have just been unnecessary weight on the way to the finish line. I was amused when White shared with me a story from his racing past, when he had the opportunity to drive a navy commander around a racetrack, much like the experience I had myself at Atlanta Motor Speedway. "When we really got going, he was hanging on so tight that the veins on his arms were sticking out," White said. He laughed, adding that even that many years ago, reaching a speed of 165 mph was not at all uncommon. "When the drive was over, the commander invited me to take a ride with the Blue Angels and see if I'd get as scared as he did in my car. I took him up on that offer a few months later, and it was great. I loved it. They even made me an honorary member of the Blue Angels after that." The thought of taking a ride with the Blue Angels being "fun" just floored me. Loved it? I couldn't imagine. While my drive at Atlanta Motor Speedway was one of the most thrilling things I've ever experienced, I can tell you unequivocally that a ride-along with the Blue Angels is not in my future. When White said that he'd conquered fear, I believed him. The smallest man ever to capture the NASCAR championship stands just five feet four inches tall and weighs about 135 pounds, but his mark on the sport looms large.

White's accounts of his early days in NASCAR painted a very clear picture for me with respect to how much the sport, the science, and the players have changed since 1947. Talk to a few crew chiefs, and you'll get very different answers as to what wins races. Some say the engine makes the difference. Some say it's the chassis, since different tracks respond better to some designs than to others. Some will swear that it's the tires that mean the difference between a win and a loss. Still others will tell you that the car is critical, but without a gutsy, resourceful, fearless driver and crew, winning a race simply can't be done. Ed Clark shared a story with me about "King" Richard Petty, one of racing's legends who won the NASCAR Championship a record seven times and won a record two hundred races in his career. Statistically, he is still the greatest driver NASCAR has ever seen. In 1964, Petty won his first race at Daytona, clinching the top spot in a powerful blue #43 Plymouth with a 451 hemi that his brother, pit crew chief Maurice Petty, modified for safety reasons. How did he do it? Maurice reinforced the interior of the car—the driver's seat, specifically—using thick foam, a couple of two-by-fours, and duct tape in order to keep his daredevil brother from flying around inside the car, especially in the turns. To a Southerner, there's pretty much nothing that can't be fixed with duct tape and a two-by-four. I'm finding that I do love that about NASCAR. It's not always the money behind the driver that molds a champion; very often it's the do-whatever-it-takes mentality. One thing I can assure you that neither White nor Petty foresaw back in the earlier days of NASCAR was the here-to-stay entrance of women onto the racing scene, either in the driver's seat or over the wall (in the pit) at a race. While Sara Christian was the first woman ever to drive in a NASCAR race, she was, in her own estimation, an unusual variety of woman. Competing for the first time ever at Charlotte Speedway in 1949, she continued to race for years after that. She and her Ford made for formidable opponents. Her best-ever finish in 1949 at Heidelberg Raceway in Pittsburgh remained the only top-five finish for a woman until 2011, when her record was broken by Danica Patrick. Make no mistake, Sara Christan, wife of NASCAR driver Frank Christian, was the exception and not the rule in the early days of NASCAR. Neither fans nor the sport were ready to accept a woman's perfumed presence in the midst of a rough and tumble, dangerous sport like stock car racing. Women have sashayed onto center stage, slipping behind the driver's wheel and jumping over the wall to work in the pit. They are not oddities or amusing side stories. They are fierce competitors who mean to win every time they strap on a helmet. You might think that their presence might be a problem for other drivers and fans alike, just as you might think that seeing a woman play linebacker on an NFL team might shake people up a bit. Surprisingly, I don't think that's the case. In my experience, the fans love seeing women out on the track, mixing it up, and not backing down. Danica Patrick is a fan favorite, and I've seen about as many men wearing her merchandise at races as I have women. I also had the opportunity to sit down and talk with rookie NASCAR Nationwide Cup driver Johanna Long, the beautiful twenty-year-old "girl-next-door" who lives—and loves—to race. As a mom and self-proclaimed chicken, I had to ask the Indiana native the question that first came to my mind when we talked: are you ever afraid out there on the track? "No, it's never fear," she answered, very matter-of-factly. "It's a rush. I don't think about anything else but doing my job when I put on my helmet." She tried ballet and softball as a child, but she was never as happy in those activities as she was when she was behind the wheel of a race car. She laughed when she remembered something that a former team member told her a few years ago. "I had an old pit crew chief tell me one time, 'Jo, you're happiest when you're racing. It's what you were born to do.'" I digested that answer for a minute, trying to grasp the concept of a woman—who is a year younger than my youngest daughter— being a NASCAR driver. Then I had to ask her the second question that came to mind once she told me that. "What did your parents say when you told them that you wanted to be a race car driver?" Her answer to that one gave me something to think about too. Her parents simply let her love of racing guide her, just as parents of a ballerina or a football player might do with their child. Johanna Long has raced competitively since she was eight years old. She's been going to races since she was even younger, as her dad raced late-model cars (another version of stock car racing) for as long as she can remember. Long raced go-karts first, then Legends cars at age twelve. Legends car racing is a form of racing designed primarily to promote exciting, heart-skipping competitions while keeping costs down. Legends car body shells are five-eighths-scale replicas of American cars from the 1930s and 1940s, and they're all powered by motorcycle engines. At age thirteen, Long began racing stock cars. She entered NASCAR by racing in the Camping World Truck Series at age eighteen. As we talked about that experience and others, she shared something with me that surprised me, saying that she does not feel that she is treated differently by the guys on the track. In fact, she feels as though she gets a lot of respect. But she's earned it, just like everyone else does. "It takes a while for other drivers to learn how you drive. Once they do, they respect you." Currently, Long only gets in about twenty-one races a year. Why? Because track time costs money, and she needs more sponsors in order to be able to race more. I'm finding that every driver, every team owner, every player in NASCAR has that same thing on their minds—sponsors, and how to get them. That question and the one of how to entice and engage a wider fan base (including a younger genera-tion of fans) are probably the two most-asked questions out there, if you're in any way connected to NASCAR. I get the feeling that Long will get the sponsor thing figured out, and that we'll be hear-ing a lot more from her in the coming years. She loves racing, and she's hungry. "I'm not the youngest driver in NASCAR, by a long shot," she said, adding that there are a lot more young women coming up the ranks behind her. Change, it seems, is coming again. Buckle up, because I do believe NASCAR is ready and waiting.

Chapter Nine
The Evolution of the Sport—the Cars

I can say without a doubt, even in my limited understanding of cars and engines and such, that the days of shade tree mechanics working on their cars to get them ready for Sunday afternoon races are long gone. I had the opportunity to tour the impressive Penske Racing facility in Mooresville, North Carolina, to learn firsthand what all the fuss is about with respect to the "shops," or garages, that house the teams and build the cars for NASCAR. It took me a few weeks to talk myself into making that trip, but I finally came to the conclusion that I could not responsibly or intelligently write about the cars and the monumental behind-the-scenes efforts without seeing the shops firsthand. Even so, I still secretly dreaded the prospect; I pictured dark, dirty concrete garages flooded with fluorescent lighting and filled with the insistent noise of air-powered tools and depressed-looking mechanics, shouting at each other. That's what the place looks like where I get my oil changed and tires rotated, anyway. I reluctantly but obediently found a shop in North Carolina and made arrangements for my husband and me to tour it.

Several big-name garages are located in North Carolina, including Roush-Fenway Racing and Hendrick Motorsports in Concord; Joe Gibbs Racing in Huntersville; and Michael Waltrip Racing in Cornelius. There are many others of course, but these are just the names I had picked up from various conversations I had earlier, with people who seemed to know what they were talking about. I understand that all the racing shops, no matter which ones you choose to tour, are really something to see. North Carolina, as it turns out, really is a hotbed of racing talent.

The Penske Racing Team shop is unobtrusively situated on 105 beautifully landscaped acres in Mooresville, not far from the NASCAR Hall of Fame. Roger Penske, former race car driver and owner of racing teams in both Indy and NASCAR, owns the sprawling facility. In the main entrance of the building, glistening Italian floor tile, black leather chairs, and large windows say a gracious hello to visitors. The place could easily pass as a high-priced attorney's office or some other such professional lair, not a place where cars are built. Unless I had known that I was there to watch mechanics work on cars, I'd have had no idea that there was a massive race car building facility just the other side of this glamorous reception area. There were no loud noises, I didn't hear any shouting, and I didn't smell oil or rubber or whatever that smell is that lingers in any mechanic shop, no matter where it is.

When we found our way around to the back of the building and entered through the guest entrance, we were met not with the smells of grease and rubber, but with two smiling women working the desk in a neat and clean gift shop. Glassed trophy cases quietly boasted that Penske's is one of the most successful teams in the history of the sport. We saw all kinds of fan merchandise for the drivers on the Penske team, for both the NASCAR and Indy Racing League. The back wall of the gift shop was glassed, giving us a view of a huge brightly lit garage. Clean, neat, uniformed mechanics were busy at work on several cars. I felt a huge sense of relief at this point of the tour; seeing all that bright light and having an opportunity to shop put me right at ease. I don't know why so many car maintenance shops are bleak, dirty, and depressing-looking, but in my opinion they are. On the other hand, the Penske Racing facility is cheerfully bright and hands-down cleaner than my house. I'm not sure whether that's a more telling commentary on the garage's cleanliness or my poor housekeeping skills, but there it is anyway.

My husband's eyes widened; they had a glassy sheen to them. He looked just like an unsupervised kid in a candy shop.

"Are you here to tour the shop?" one woman asked, and when we said that yes, we were, she directed us to the stairs behind her desk that led to the Fan Walk. The Fan Walk is a railed walkway overlooking all the action in the shop below. No appointment is necessary to tour these garages; you can simply show up and spend as much time as you like watching the mechanics do their thing. I looked from my husband to the stairs and back again, sighed a deep, "This is going to be boring," sigh, and started climbing.

The view from the 432-foot Fan Walk is all-inclusive, save for the goings-on behind the walls that shield the far half of the nearly 425,000-square-foot shop from curious eyes. All the walls are neatly lined with red toolboxes, something my husband noticed right away and admired aloud as we walked. Right away, he loved this place. He, and many men I know, feel as passionately about red toolboxes stuffed with tools as I do about fine Italian leather handbags. You can never have enough of them.

The mechanics we saw were busily working on a row of cars, all of them numbered either #2 (Keselowksi), #22 (Joey Logano), or #12 (Sam Hornish). A separate area of the shop is dedicated to building the Indy cars. While we didn't get to see any that day, posters displayed all along the Fan Walk featured both the cars and their drivers. Even for someone who is not a car aficionado, I can say that both the NASCAR and Indy cars are beautiful things, works of art. Each numbered car we saw looked exactly like the others in the shop that sported the same number. Every #2 car was blue, every #22 car was red and yellow, and every #12 car was black and yellow. It's kind of embarrassing for me to admit this now, but that day at the Penske Racing shop was the first time it dawned on me that every driver has more than one car; rather, every team builds more than one car for the driver to race. I suppose I actually thought that a driver had one car, drove it in a race, and if it got banged up, they fixed it before the next weekend. That obviously makes no sense, I know, but I meant it when I said I knew absolutely nothing about NASCAR when I started this journey.

One of Brad Keselowski's race cars being worked on in Penske Garage.

One of the mechanics who works on Sam Horhish's Nationwide series car, Kyle Pinkerton, was kind enough to take some time out of his busy day to show us around and explain a few things to me. For instance, I noticed while we watched the mechanics working that there are three holes in the back window of every car. Two of those holes are rimmed with neon orange, and every now and then, a mechanic would stick a tool into one of those holes and make some sort of adjustment, looking underneath the car. I couldn't imagine what those holes were for (why on earth would anyone put holes in a window?) and how they could possibly affect something down by the tires. Kyle patiently explained, then demonstrated for me, what the mechanics were doing when they made those adjustments.

It's actually very interesting, as are most of the magic tricks performed so flawlessly by pit crews during races. When a driver pits his car at a race and the pit crew rushes out to do their jobs, one crew member will often make what's called a "wedge adjustment" using a special wrench fitted into those orange-rimmed holes. In the movie *Days of Thunder*, which I've watched all the way through twice now, actor Robert Duvall referred to this action as "setting up the car." This adjustment changes the amount of tension on a spring in the rear suspension of the car. It can be made in a matter of seconds, using those ingenious little holes in the window and a ratchet, also known as a socket wrench (I asked). The holes, interestingly, are marked with neon orange so that the mechanic can see them right away and not waste any time looking for them during a pit stop. Pretty smart, I thought, but even I had to ask the question, "Why make that adjustment during a race if the mechanics are obviously able to do it right here? Doesn't that waste time?" Kyle, who obviously loves what he does and seemed to enjoy helping me understand it, explained it to me this way: A wedge adjustment can make a dramatic difference in how a race car handles during a race. The amount of tension needed on the springs can vary from track to track, and it can even fluctuate during the course of a race. The suspension system is directly affected by changing track conditions; it probably won't be exactly right until the driver feels, then communicates to his team, how the car is handling. At that point, during a pit stop, the wedge adjustment will be made. Those precious seconds and steps would not be wasted unless the adjustment was a critical one. As Kyle explained all this, we got down on the floor right there in the shop and looked underneath one of the race cars. He showed me the springs, then demonstrated how the adjustment is made. All the time that I listened to his explanation (which was very good, by the way, because I actually understood it), I thought to myself, *Suspension? You mean with everything else going on in a race, the driver cares whether his ride is bumpy?* The few times my own mechanic back in Georgia ever talks to me about suspension is when I am complaining of a bumpy ride in my otherwise cushy suburban mom-car. In fact, he talks to my husband about it, not me.

Anyway, the puzzled look on my face gave away my confusion, I suppose, because Kyle went further to explain how the suspension can make all the difference between winning and losing. It's not about the comfort of the ride. It's about the efficiency. Springs on all four tires of a car combat the wasted energy that bumps cause on the track. When the tires hit a bump (no matter how small it may be), the springs use their force to push back, instead of wasting the energy of the car absorbing the bump. Proper suspension makes the most of the car's forward motion, and it also ensures the best possible grip that those tires have on the track. In other words, the suspension can mean the difference of a few precious seconds, the difference between crossing the finish line first, second, or dead last. Look closely at the back window of a race car the next time you see one. Then you can impress your friends by explaining why those holes are there. I do, every chance I get.

The spotless tile floors and bright overhead lights in the Penske garage are the backdrop for an operation that designs every car from the ground up, literally. The frames are built from heated and bent raw steel. Chassis are built. There is a fabricating department, a separate area for gears and transmissions, and a room dedicated just to tires (always eleven inches wide, never any tread). There is even a practice pit in which an electric car is used, eliminating the danger of toxic fumes in the workspace. Crew members can work on their time and technique, shaving off a tenth of a second here, another there. Everything in NASCAR, it seems, is about speed.

Two shiny eighteen-wheeler trucks, one that houses and carries Logano's car and one for Keselowski's, were backed into the shop that day, both of them open and ready to swallow up and transport the primary and backup cars for both drivers to the next race. The walls of the huge trucks are lined with toolboxes that contain every part and tool that could conceivably be needed during qualifying races and the actual races. A hydraulic lift raises one car to the top level of the trailer, leaving room for the other car on the floor of the trailer. A kitchen and a lounge are also housed in these impossibly clean and organized trucks, though the lounge is typically used for team meetings, not relaxing and watching television.

Inside the Penske Racing garage, view from Fan Walk.

There are about one hundred Indy engineers and mechanics and about three hundred NASCAR engineers and mechanics on site. Each team can have as many as thirty engineers on it, and there are also some very talented mechanics on those same teams.

The shops in North Carolina, also referred to as Racing Country, USA, are impressive sights to see. While I did leave Penske Racing feeling somewhat inadequate with respect to my housekeeping skills, I thoroughly enjoyed the experience. I highly recommend making the pilgrimage to Mooresville to see these marvelous displays of perfection and engineering brilliance.

I have learned during this journey that the cars of NASCAR are now in their sixth generation, at least in Sprint Cup racing.

One of the bare chassis constructed at Penske Racing.

As impressive as the cars of today's NASCAR are, I don't think their strength and engineering can truly be appreciated until the big picture of the history and evolution of the cars is understood. I could dedicate an entire book to this subject alone, outlining the changes made along the way and over almost seventy years, in the interest of both speed and safety. However, I've narrowed the topic to focus on the most recent two generations of cars, since they're both still competing on tracks all over the country today.

The fifth generation of NASCAR machines was called the Car of Tomorrow. The product of about seven years' worth of research and design, the Car of Tomorrow was born from the tragic death of Dale Earnhardt during the 2001 Daytona 500. According to Kyle, who has continued to answer questions for me long since crawling around on that floor months ago back in Mooresville while trying to explain wedge adjustments, NASCAR first started using the Car of Tomorrow in Sprint Cup racing in 2007. It wasn't used in the Nationwide Series until 2010, and it is still the car raced in that series today. Reviews of this iteration of cars have been mixed, but I suppose that's true of every new generation of car in the sport. Change is never easy, is it? Drivers and fans alike have said that a problem with the Cars of Tomorrow is that they really don't look like their counterparts out there on the street. For example, a COT Ford Fusion doesn't look much like the Fusion that rolled off the manufacturer line, the one that the fans drive. Neither does the Camry or any of the cars of that generation. Cars of Tomorrow are somewhat boxy-looking and were designed to be safer, to be less expensive to maintain, and to encourage closer competition. You may think that the actual appearance of the NASCAR race cars is no big deal, but it's a huge deal to fans. Remember, NASCAR fans love a particular car manufacturer. No matter who that manufacturer is, the fans want to see cars on the track that look like they're *supposed to* look. A Camry on the track should look like a Camry in the supermarket parking lot or on the highway. It's just an unwritten rule, part of the NASCAR fan code.

In a nutshell, in this fifth-generation car, the driver's seat was moved four inches toward the center and the roll cage three inches toward the back of the car, both moves designed to enhance safety. The Car of Tomorrow has larger crumple zones than its predecessor and is taller and wider. The exhaust was moved from the left side of the car to the right, channeling heat away from the driver. The fuel cell is more powerful and has a smaller capacity.

There is something I learned about this car, and I found it fascinating. The Car of Tomorrow enabled the debut of a racing phenomenon called lock-bumper super-drafts, in which two cars can bump, then actually lock together. Since the body styles of all the Cars of Tomorrow are pretty much alike with their bumpers the same height and width, this bumping action doesn't typically have the expected effect of upending the bump-ee and causing a terrible crash. Still, it's a nerve-wracking thing to witness for someone like me, who maintains at least ten car lengths between my car and the one in front of me on the highway. This gutsy move can help both drivers achieve speeds up to 10 mph faster than those who have a conventional draft (without bumpers locking), sometimes topping 206 mph while locked together. Based on what I've read, this racing phenomenon first occurred in 2009 at Talladega Superspeedway, and in 2010 fans were treated to the exciting spectacle of a previously-unheard-of 175 lead changes on that same track. In 2011, three or four groups of super-drafters would break away from a pack of cars racing in a tight bunch on the track. These groups, or tandems, would generate an effect called aero-push, in which not only were the participating cars drafting, but the air push they created by doing so actually hindered the cars behind them from passing. Aero-push is a technique often seen in Formula One racing, and it encourages teamwork among drivers. All of this stuff is science, which can be traced back to Sir Isaac Newton's three laws of motion. These guys aren't dummies, either.

As EXCITING AS ALL THIS SOUNDS (and it really is thrilling to witness), racing great Dale Earnhardt Jr. and veteran drivers including Richard Petty and David Pearson have been very outspoken about their dislike of this style of racing. In fact, I believe one of the drivers actually called the technique "crap." A record sixteen caution flags were waved in one single race, most of them being thrown when pushing cars actually spun out the leader of the pack. While many race fans will say, "Leave them alone. Let the guys race," some drivers and many more fans, along with NASCAR's movers and shakers said, "Enough." In 2012, a curved spoiler and lower, longer rear bumper were mandated, putting an end to the breathtaking phenomenon of the lock-bumper super-draft. Well, mostly anyway. I'm learning that no matter how the cars are changed, even with safety at the center of those changes, fearless drivers will figure out a way to push the limits. Now, let's take a look at the Gen 6 cars, the ones initially tested and driven during the 2013 Sprint Cup race season. As with the Cars of Tomorrow, these machines are a fluid work-in-progress. In a May 25, 2013, NASCAR media teleconference, Brian France talked about the sport's dedication to both research and safety, as well as enhancing racing excitement. "We'll be going in a direction that I've told everybody ... we're going to use a lot more science than art in establishing the very thing that matters most, which is safety. A central goal of NASCAR is to obviously have safe racing and at the same time have the tightest, closest races in the world. That's our mission," France said. France is pretty clear on the subject, and the Gen 6 car is supposed to be even safer than the Car of Tomorrow. The window nets have been redesigned to keep stray debris out and the driver's limbs in, and a roll cage bar has been added. The windshield is now a laminate, so it's more resistant to shattering when brake parts or other debris hit it. Other safety measures have also been taken.

Driver Brad Keselowski (#2), in a March 2013 blog he wrote for *Autoweek*, said that he definitely sees more speed in this newest iteration of NASCAR design. For any driver, I will assume that's a plus. Driver Marcos Ambrose had this to say about the new design: "This is a great car. It's really light and extremely fast. I think we're seeing faster speeds each week." With regard to the Gen 6 car's fan appeal, Ambrose added, "It's more similar to the look of the actual factory cars, and I think that's been good for fans. They can recognize the car they drive in the ones I race on the track." But in a story that flew through the ranks in NASCAR and eventually erupted into mainstream publications such as the *New York Times*, driver Denny Hamlin (#11) openly criticized the Gen 6 cars. His complaint? He compared the new car to the past generation of cars and said, "I'm never going to believe in it." After racing his Gen 6 car in Phoenix, Hamlin told reporters, "I don't want to be the pessimist, but it did not race as good as our Gen-5 cars. . . . This is more like what the Generation 5 was at the beginning. The teams hadn't figured out how to get the aero-balance right" (*ESPN NASCAR*). Three days after Hamlin stated his opinion, NASCAR fined the driver twenty-five-thousand dollars for publicly airing his "disparaging remarks." Brian France and the sanctioning body have a vested interest in and keen sensitivity to the fans' perception of this new generation of cars, and they did not take kindly to Hamlin's remarks. Many fans thought the fine was too steep a penalty for Hamlin openly expressing his opinion, but the disciplinary action stood. Cooler heads eventually prevailed; Hamlin did pay the fine and said that he was moving on with the 2013 season and his pursuit of the championship. NASCAR definitely has a reputation for sending very clear, expensive messages when they want to emphasize a point, especially when it comes to conforming to policy where the cars are concerned. For example, NASCAR came down very hard on drivers Brad Keselowski and Joey Logano and their teams during the 2013 season of testing and implementing the Gen 6 cars. Charging that both Penske Racing teams brought unapproved rear housing parts to an April race at Texas Motor Speedway (and modified their cars with those parts), NASCAR dealt hefty fines and penalties across the board: among the hand-slaps were six-race suspensions for seven crew members of both Brad Keselowski and teammate Joey Logano. Added to those suspensions were two hundred thousand dollars in fines. The drivers and Roger Penske were each docked twenty-five championship points, a swift move that dropped Keselowski from second to fourth in the Sprint Cup standings, and Logano from ninth to fourteenth. Each of the drivers' crew chiefs was fined one hundred thousand dollars suspended for the next six championship points races, and placed on probation for the rest of the calendar year. The team manager for both cars was suspended and put on probation, and the individual car chiefs and team engineers were also disciplined. NASCAR confiscated the parts. The disciplinary actions were upheld in a 2013 appeal, but Penske said he would again appeal to NASCAR in a last-ditch effort.

In that same race, driver Martin Truex Jr. (#56), who finished second at Texas behind Kyle Busch, was punished for the front end on his car being too low. He and owner Michael Waltrip were each docked six points, and crew chief Chad Johnston was placed on probation and fined twenty-five thousand dollars. Of course drivers, owners, and crew members can always appeal any disciplinary action, but the initial message is clear: teams must conform to the rules, or pay the price.

The evolution of the cars in NASCAR has been purposeful, with every decision being made for a reason. That reason may be in the interest of safety or to enhance the fan experience. Recent changes have even been made to the cars in order to increase manufacturer brand identity because, believe it or not, successful cars on a NASCAR track result in more sales on the dealership floor. Who knows what changes will come with the next generation of cars? There will definitely be a next generation, and probably another and another. As the drivers and their teams continually hone their crafts, so must the cars keep pace with them.

Chapter Ten

"This Is Talladega!"

Having been to the race in Daytona, having watched seven or eight races on television and taken copious notes about them, and having done a few months' research since the February Daytona 500, I decided I was ready to attend another NASCAR race. This time, I'd know a whole lot more about what I was watching. By the time I had definitely made up my mind to go to another race, it was near the end of April, so the only logical choice was to go to The Big One, Talladega Superspeedway in Alabama. This racetrack—this race, in fact—is a legend among NASCAR fans. Talladega Super-speedway is a massive 2.66-mile, 33-degree banked track, well known for its super-wrecks, aptly named the "Big Ones." Because of the banking and widths in the turns, cars can race three and four wide, and one mistake, one wrong move, can be disastrous. It's not a matter of if a wreck will happen; rather, it's a matter of when. Talladega is one of the NASCAR tracks at which car engines must have the restrictor plates installed on them, the ones that harness horsepower and rpm's in the interest of driver and fan safety (the restrictor plate is also that shiny thing that doubles as a great com-pact mirror). The reason for reigning in the power that those cars can churn out is that, on the track's straightaways, today's super-cars could easily reach 225 mph. A slight nudge or bump in a tight pack of cars, a blown tire, or a tiny speck of grease on the track, could spell disaster for several drivers and possible fans on the sweeping Talladega track. We bought our tickets (I charmed my husband and son into going to this race with me) and our passes for what's called "The Talladega Experience," and eagerly looked forward to May 5, the date that the Sprint Cup race was to be run. I was more excited about going to this second race than I expected to be. I was feeling the first stirrings of becoming a true NASCAR fan.

The event in Talladega lasts a full three days, with qualifiers to be run on Friday and Saturday, and the Nationwide Cup race to be run Saturday afternoon. The typical weather in early May in the South is about as close to perfect as it gets, with warm, breezy days, low humidity, and rare rain showers, just enough to keep May's flowers bright and beautiful. In 2013, however, Mother Nature had different plans. As soon as we bought our tickets ($55.00 each, the least expensive ones available) and our three "Talladega Experience" passes ($145.00 each, entitling us to all the barbecue we could eat and all the alcohol we could drink before noon, a pre-race tour of the pits, a driver Q&A session, and proximity to the drivers when they were introduced—a bargain, really), I whipped out my handy race day checklist and began to go through the items one by one, making sure my race day bag was properly packed. Sunscreen—check. Binoculars—check. Money, lip gloss, fun but not obnoxious fan paraphernalia—check. I also learned at Daytona that fans are allowed to bring into the track coolers with their own food and alcohol. What other sport lets fans do that? That really surprised me, but I remembered it, so I put on my June Cleaver apron, and I packed a lunch cooler that would have made Beaver's mom so very proud. I packed turkey sandwiches on whole wheat bread, fresh fruit, popcorn, crackers, and sliced cheese. I packed a separate cooler with water and sports drinks. At this stage of my journey, I think one could officially call me a "NASCAR Nerd" and be right.

I considered this to be my first mini-excursion into tailgating, because the awful, rainy weather in Alabama did not encourage full-blown tailgating as I had seen it in Daytona. Besides, there were camping areas at Talladega I wouldn't dare take a stroll through, no matter who was cooking what. Even the infield campers, supposed to be the upper echelon of race partiers, looked a little rough to me. Mind you, these eyes were at the time still very inexperienced; I'm just telling you what I thought. There were some nice RVs, yes, but many people had constructed shaky scaffolding on which to put their chairs to watch the race, and one guy right across from where we were sitting watched the entire race shirtless, wearing only a pair of jeans. He did not have the physique to be doing that, but he did it anyway. He was drinking straight from a bottle of Jack Daniels whiskey too, and on my mother's grave, I could hear him belch occasionally from all the way across the track and the forty-three cars racing around it. The scaffolding held, and I suppose that's all that matters, but no, I did not wander out there to socialize with people as I did in Daytona. I'd been told that every race has its own personality, and some of the things I saw being played out at Talladega certainly reinforced that race's reputation. My girlfriend's words, the woman who so adores NASCAR and who can cite facts and statistics like nobody's business, kept echoing quietly in the back of my mind, "You'll be fine as long as you're out of there by dark. . . ." Her words gave me chills.

As soon as we had plopped down a cool $650 or so for three "cheap seats" race tickets and three "Talladega Experience" passes, Doppler radar screens all over Alabama went solid green, with a few yellow and red splotches thrown in for fun. Weather screens in both Georgia and Mississippi did the same thing. In fact, before my husband hung up the phone after purchasing the tickets and passes, giant black and gray rain clouds from who-knows-where rushed down here with reinforcements and settled comfortably in for a week, over Mississippi, Tennessee, Alabama, and Georgia. An entire week. That almost never happens in May in the South. What the gracious South typically serves up in May are crystal blue skies, lacy white dogwood trees, pink, airy azaleas, and stately magnolia trees, gracefully draped in their spring finery—glossy green leaves dotted with beautiful pearly baubles, the unmistakably lovely magnolia blossoms. We have warm days with cool, breezy evenings, perfect for sitting on the front porch with neighbors and sharing a glass of sugary iced tea as only a Southerner can make it. What we most certainly do not expect in early May is a monsoon.

Another little known fact, or more precisely, one I had forgotten, is that NASCAR races won't be run in the rain or if the track is at all wet. Period. Well, OK, yes, I did know that, but what I found out the hard way is that, if the race for which you've purchased tickets gets rained out or cancelled for any reason, the ticket and pass prices are nonrefundable. I choke on those two words. I choke on the phrases "nonrefundable" and "all sales final," which I suppose both mean the same thing. So, from the second the ticket transaction cleared our bank, I stayed obsessively glued to the television watching the weather, as if by doing so, God would have pity on me and part the green Doppler echoes just for Sunday, just for the big race. Some friends of ours, NASCAR fans from way back, had to cancel their plans to go with us because of the threatening weather, so that also kicked my stress up a notch. I have heard horror stories about the happenings at the Talladega race for years. I know that sounds bad, but the people telling me the stories were NASCAR fans. Praying that those stories were just Alabama's version of urban legends, I called my girlfriend in a panic and confessed that I was afraid to go to Talladega without her and her husband, longtime pros who know the ropes. Laughing, she assured me that we'd be OK, as long as we were "out of there by dark." Those words again. I felt a tight, hard knot forming in the pit of my stomach. "Why?" I asked, in a shaky whisper. "Well you know," she answered, "people will have been drinking pretty much constantly since Thursday or Friday, so they get a little rowdy if their guy doesn't win. They get really rowdy if their guy does." That, to me, sounded like a lose-lose proposition for a newbie racing fan. "Oh," was all I could muster in return.

After she and I had that conversation on the Saturday before the big race day, I turned on the Weather Channel and sat glued to it for the rest of the day. Even if the race date got pushed back to Christmas Day, we were going to go. But for now, we were planning on Sunday, May 5. At 5:00 a.m. Sunday morning, we loaded up our nerdy race coolers and my race day bag and headed west. I slept almost the entire two-and-a-half-hour ride, with my husband waking me up about fifteen minutes outside of Talladega. It had stopped raining; in fact, the sun was trying its best to peek through the dense clouds, giving millions of optimistic race fans hope. As the darkness was lifting and the fog of deep sleep was clearing from my eyes, what was coming into focus on both sides of I-20 was a flashback I remembered from a trip I took many years ago to Haiti. While Haiti is blessed with some of the most beautiful scenery I've ever seen, it's also one of the poorest countries I've ever visited. The Alabama countryside surrounding Talladega Superspeedway was dotted and splotched with replicas of the makeshift housing I saw in Haiti, which consisted of tents and colorful but dilapidated boxes, really, connected mostly by clotheslines with filthy clothes hanging on them. Here in Talladega though, this colorful, dilapidated neighborhood had been hastily slapped together by drunk contractors doubling as race fans. In some areas, there were tents as far as the eye could see, and most of them appeared to be flooded. In other areas, old broken-down campers were sucked hubcap-deep into that unmistakably red, wet Alabama clay, kind of a very thick yet slick quicksand, and sure death to tire traction. It had been raining all weekend, and port-a-john outhouses were leaning precariously to the left or the right, threatening to spill their secrets out onto the tents next to them or downhill from them. My stomach rolled at the thought. The weather had obviously not been kind to these camper- and tent-dwellers, and as we exited the interstate and drove closer, then through, these villages, I instinctively locked my doors and stared straight ahead. "Look, there's a pink school bus, Mom!" my son exclaimed. "Hey, what are they smoking?" I stared straight ahead while my husband deciphered directional signs. We, too, were driving in Alabama mud, trying to find where to park. I kept checking the skies for clouds, but they seemed to be parting, not gathering again and knitting together. It was looking like we'd be going to a race that day.

If it had been any other race of the season (and if we hadn't already paid a lot for these tickets and passes), I would have simply said, "Forget it. We'll go to another race instead." But this was Talladega. Skipping it would be like going to New York City for a visit and skipping a foray into Manhattan. It simply isn't done.

By the time we parked our car, it was about 8:00 a.m. We unloaded our neatly packed coolers and backpacks (that was another smart change I made since Daytona—a hands-free race day bag), and started trudging toward the huge stadium just ahead of us. But between us and the racetrack, there stood a carnival of sorts, with vendors hawking everything from scanner headphones to smokeless tobacco to timeshare condos in Vegas. One large sign admonished race-goers to brush their teeth twice a day, two minutes at a time (sound familiar?). Another one stated, "No hair—cute, but no teeth—not cute." Flossie the packet of dental floss was there too, passing out toothbrushes and rolls of the thin white string. In fact, on the way back to Atlanta later that evening, those same two messages about proper brushing were displayed on billboards high above Interstate 20. Rule No. 1 in marketing is that you only advertise where you know there's a market.

At 8:00 in the morning, there were booths set up and open for business, with shingles hung out advertising BEER, WINE, and COCKTAILS. They had been selling, and people had been buying, since about 7:00 a.m. The Copenhagen smokeless tobacco tent already had a line of people waiting to get in that wrapped around the building at least twice. For some reason, selling alcohol by the drink to fans that early in the morning struck me as a bit odd, but as my husband explained, many of those people had been drinking straight through since Thursday. It wouldn't have mattered if it was 8:00 in the morning or 11:00 in the evening. Clearly, if I do decide to become a full-on NASCAR fan, I will have to train hard to reach that level of drinking prowess.

Before we reached the edge of what I now fondly call "carny alley," I had gotten two shouts of "Show us yer boobs" (which I did not) coming from race fans who appeared to have been sleeping in their cars and trucks. I had a free beer shoved into my hand by a passing fan, but I threw it away in the nearest trash can. Not my brand.

Once we had run the minefield of carny alley (I registered to receive information about everything from timeshares to chewing tobacco to motor oil, just to get free promotional items) and had made it safely to the gates of the Speedway, our bags were quickly checked, our hands stamped, and our pit passes and hospitality suite passes authorized. The nice man at the gate advised us to head down to pit road first, as the crowds build down there pretty quickly, so that's just what we did. The closer we got to pit road, where a few of the cars were already lined up and waiting, I began to feel that familiar excitement again. The third-world-looking circus we saw as we approached the Speedway was forgotten, as soon as the sun glinted off those beautifully muscled cars, resting now but ready to roar to life with one slight motion from their masters. All three of us thoroughly enjoyed going from one driver's pit to the next, and I remembered the lessons and facts I had learned at the shops and at the Hall of Fame in Charlotte. Pit crew members were busy at work taking care of their own area, their own job, as if there was no one else on the planet. Occasionally a fan would hold out a program or hat and ask for an autograph, and I don't think I saw one mechanic or chief refuse the request. Jovial calls of "Roll Tide" were exchanged between crew members and fans, Alabama's way of saying, "How are you?" and "I love Alabama football." My son

The markings on the Goodyear Racing Eagle tires are there for a purpose. Tracks and conditions determine how mechanics use those marks.

was fascinated with some of the preparatory procedures and could have spent hours just watching those guys work, but after a while, all I could think about was the fact that we each had a $145 pass that was supposed to entitle us to all the barbecue and beer or wine we could eat and drink right up until race time at noon. That's a lot of barbecue and beer to ingest before midday, but hey, who was I to argue? I was the new one at this. Eventually, then, we found our way to the hospitality tents. There was a band playing, there were tables and umbrellas set up throughout the area, and inside the tents, displayed on long banquet tables, was pork heaven, if there is such a place. Heaping, steaming mounds of barbecue ribs were lined up one after another on draped banquet tables. Collard greens swimming in juice and flavored with ham were displayed in heated serving dishes. Still other grand bowls offered up Southern-style potato salad. What made it Southern-style? It had everything in it but the kitchen sink, most of which I could identify: eggs, mustard, mayonnaise, potatoes cooked to just the right consistency, celery seed, celery, and a few other ingredients that made the whole mess taste just fabulous. I loaded up my plate with ribs and all the sides, opted for a glass of water (it was so early!), and found a table outside under an umbrella not too far from the stage where the band played. Drivers Brad Keselowski and Dale Earnhardt Jr. would soon appear for a short Q&A session with the fans on that same stage, and I wanted to be sure to hear that. My son and husband followed after a while. Apparently the holdup had been my son, wanting to spend a little extra time with the "Miller Lite" girls, who were ooh-ing and ahhh-ing over him, draping beads around his neck, and giving him beer. It was a twenty-three-year-old male's dream-come-true. My husband stayed back, I suppose, to supervise the whole affair. Too much of that kind of activity and my son would have likely asked one or both of the girls to marry him. When they joined me outside and I saw their plates, I felt a little embarrassed. My husband had two ribs and a bottle of water (a Yankee from way back, he hates those Southern sides, and he doesn't drink any alcohol at all), and my son had a few slices of barbecue chicken (he doesn't care for ribs—I assume he got that from his

Dale Earnhardt Jr. and Brad Keselowski Q&A session in Hospitality Area, before Talladega race.

father's side of the family). I sat there with a plate heaped up like the King of England getting ready to dig into a feast with his fingers, and those two had prepared plates that made them look like they were on day one of a very strict diet. None of us had had any alcohol at that point, as it was only about 10 a.m. Many others, though, had a beer in each hand and a couple stuffed in their pockets in case they ran out before making another trip through the line.

It wasn't until we got settled in at our little table that I had the chance to look around and really appreciate what a great place to people-watch the Talladega race is. Sure, Daytona was fun too, but there was something about this crowd that was, I don't know, a strange combination of interesting, weird, and just plain unsettling. While we enjoyed our meal, three women came prancing up to the stage dressed like Wild West hookers and acting as though they were each vying for the enviable position of "Miss Sprint Cup" herself, a spot that's already taken by a beautiful blonde, by the way. They were wearing short, short dresses, and oversized cheetah print cowboy hats and cowboy boots, and they were tanned to the point that, should they pass away any time soon, their families would need

dental records to identify them. They were already quite inebriated
at that early hour, but what really amused me was the fact that
they flirted with every man in the place, trying to get them to buy
rounds of drinks. If I were a nicer woman, I'd have clued them in
to the fact that since they had found their way into the hospitality
area, the drinks were free anyway. There were a few good sports in
the crowd though, men who were happy enough to duck inside
and bring the women free beer every fifteen minutes or so, and
each time they did, the "girls" (who looked to be about my age
under the orange skin and pale pink lipstick) squealed with delight.
Again, I would have told them about the free alcohol, but why stop
the show? The fun lasted for a few minutes longer, then the taller
woman, the one whose orange and black thong winked at us every
time the wind blew, leaned over and barfed all over the hood of
the shiny blue #2 car displayed in the area. A cleanup crew mate-
rialized, efficiently cleaned up the mess, and within five minutes it
looked as though nothing had happened. The woman continued to
giggle and hustle free alcohol, and we just packed up our stuff and
moved to another, cleaner, table. While sitting at that table, we met
two of the nicest men, Uncle Bruce and his nephew, both disabled
veterans who told us that this race was Uncle Bruce's fifty-fifth at
Talladega. They had traveled in from Tennessee and had been camping in a tent since Thursday evening. By Friday evening, there was
a foot of water in their tent. "What on earth did you do?" I asked,
horrified at the thought of all that clothing and gear being ruined.
"I don't know, we'll worry about that after the race," Bruce told me.
When Tim, the nephew, told us that the fuel pump went out on
their truck the night before, I asked again, "Oh my gosh, what did
you do?" Tim nonchalantly answered, "I don't know. We'll worry
about it after the race. We ain't missin' Talladega." And he grinned.

When 10:30 or so rolled around, I went inside and asked the
bartender for a glass of merlot. I know how that sounds, but I had
a pass to use up. Thankfully, she suggested a cheery mimosa instead,
so that's what I took. Somehow, sipping a mimosa at 10:30 in the
morning is acceptable. Sipping a rich, peppery merlot at that same
hour seems, well, indicative of a problem.

We met some really nice people in our hour or so in the hospitality area. We met some college students, young families with children in tow, older people, and couples who were there in Talladega that day to see some serious racing, a real "throw-down," as a true Alabamian might say.

As 11:15 or so rolled around, we were hustled down to an area where those of us with "Driver Introduction" passes were gathering. This group would be allowed to stand at the spot where drivers enter the track and are introduced to fans. I must admit, that part of the Talladega Experience was very cool, to see the drivers' faces, to see them walk in with their little children and sometimes their wives. I liked that part a lot, and so did my husband and son. Once that ceremonial matter was over, we went to find our seats. That is where the real show began.

Everyone who knows anything about NASCAR races will tell you (eventually), "Whatever you do, do *not* sit down low in the stands, and do *not* take the seats in the first row, under any circumstances, for any reason, ever." So, as we made our way down the steps to the seat section and numbers our tickets indicated, we stopped, of course, on the very front row, the seats right at the wall on Turn Four, and right in the spot where a car would land on a row of unsuspecting fans if it breached the fence above the wall. In fact, if my husband had told the helpful lady who sold him our tickets on the phone that we'd like to sit in the seats in which we'd be most likely to get crushed by a flying car, our seats were exactly where she would have placed us. I know that those fences are reinforced with some super-strong airplane metal or something, but when that's all that's between you and forty-three flying bombs that weigh about 3,300 pounds each, blowing past you at 190-plus mph, that's not very comforting. Remember what happened at Daytona? No matter; in for a penny, in for a pound. I was not about to go occupy someone else's seats and hope no one showed up to claim them. All the men and many of the women at this race looked as if they could pound me into the ground without much effort or provocation, so we stayed put.

The prerace festivities were winding down, and up on the megascreen in the infield popped the image of the general manager of the 175,000-seat Talladega Superspeedway, Mr. Grant Lynch, to welcome and address the crowd. He shouted into the microphone, "This is more than a race," and the crowd answered, "This is Talladega!" Lynch did that back and forth with the fans three times, and the excitement kicked up another notch or two every time he did it. The energy was palpable.

Chipper Jones, the beloved and legendary Atlanta Braves third baseman who had just retired after playing nineteen years, his whole major league career with the Braves, was the Grand Marshall of the race that day. People in these parts love that man just about as much as Catholics love the Pope. The drivers, one at a time upon introduction, each made a lap around the track, standing in the back of a pickup truck and waving as part of their prerace introductions. The crowd rewarded their efforts with reactions ranging from adoring applause to thumbs-down "boos" and snappy one-finger salutes. *Oh my!* I thought, as I tried to ignore the rude gestures.

Chipper belted into the microphone, "Gentlemen, start your engines!" And oh, did they ever.

I do believe I am becoming addicted to that initial guttural growl that only these engines can make when they're awakened.

Danica Patrick before Talladega race (driver intro laps).

Yes, all forty-three drivers started their engines at the same time (my knees get weak just thinking about it), and then they revved them a few times for the crowd, who cheered their approval in response. Then the procession began, something akin to a slow, winding tease, with the cars purring as a tiger might, but obediently holding back behind a pace car, which on that day was driven by beloved University of Alabama quarterback AJ McCarron (another athlete adored by fans around here). Driver Carl Edwards (#99, and the guy who does a backflip out his window when he wins) was on the pole at that race on May 5, and behind the pace car, he led the other forty-two cars around the track three times before the green flag dropped. At that moment it sounded as though the mechanical hounds of hell had been unleashed. The first time that pack of powerful stock rockets blasted by us on Turn Four, seemingly inches (but in reality, several feet) from us, I have never heard, felt, or smelled anything like it. The noise was even louder than what I experienced in Daytona, because I was sitting higher up at that first race. The smells of fuel and rubber were there again, just as I remembered from that race back in February. And every time the cars blasted past us, a residue of sand and tiny flecks of rubber coated our skin,

Talladega race pace car with AJ McCarron (U of AL QB) driving.

When a race really gets underway, the sound is deafening (shot at Atlanta Motor Speedway).

hair, and clothes. This time, however, I was wearing black, not white. I was, in fact, wearing a black Clint Bowyer T-shirt I bought just as we entered the racetrack and the sun really popped out and started glaring down on the thousands of fans perusing the merchandise. I chose a black shirt, of course, having painfully relived the sickening slow-motion mental movie I had of throwing my white linen suit in the trash when it came back from the cleaners after the Daytona race. It was ruined. Nicole Kidman either wore her white suit in just that one scene in *Days of Thunder*, or she simply has a better dry cleaner than I do. Nevertheless, black is very slimming, and I like wearing merchandise endorsed by a driver who seems to find himself in the top ten or so every week. "I feel like I'm on track," Bowyer told me in a brief interview midseason, "to be a serious contender for the Sprint Cup. We just have to keep running on eight cylinders, keep working hard, and I feel good about it."

Once the excitement of the first few full-speed passes was over, most fans sat back in their seats to settle in for the next 180 or so laps. The Aaron's 499, which is what this race was called, is a 188-lap free-for-all around the 2.66 mile track. I haven't done the math myself, but I trust they checked it before advertising the race everywhere.

Here's where I really got to see where the Talladega fans get their unparalleled reputation. Once my husband, son, and I got situated in our horribly dangerous front row seats, and it was clear to everyone around us that those were, in fact, our seats, here came two men right past us with their soft-sided beer coolers, standing right in front of us, obviously planning to stay a while, maybe even for three or so hours. They were entirely oblivious to the fact that there were spectators sitting right behind them, but I'm not sure they would have cared if they had been aware. From the looks of those two, I'd say they started drinking about five days earlier; from the smell of them, I'd say they'd been camping for that same amount of time, as well. The older one was tall, thin, and leather-tanned. He had five or six teeth in his mouth. The younger one was shorter, not as thin, but his teeth would soon be going the way of his buddy's, that much was a given. And there they stood, drinking their canned beers shoved down into their coozies with their favorite drivers' numbers on them (#48 and #88) as if they were the only two fans in that entire 175,000-seat arena.

I looked from those two (I nicknamed them "The Debate Team" just because the thought was so ridiculous) to my husband and back again, thinking, *Surely those two are not going to stand there in front of us throughout this whole race.* They were, in fact, leaning against a large metal sign anchored low to the ground that read, "Please do not stand along the rail." I guess they weren't big readers, either.

In hindsight, I'm kind of glad those two stood there, or close to there, throughout most of the race. As exciting as racing can be, I still found my attention wandering every now and then, and these two gentlemen and their entourage were quite entertaining. Their women joined them after about an hour, although they sat in some-one else's seats and didn't block any more of the view. I can refer to the ladies as their "women," because that's how these men referred to them: "Woman, gimme one o' them cigarettes." "I done told you once woman, we ain't goin' to see yer mama next week, and that's that." Those are quotes, verbatim.

What really fascinated me about those two, though, was the quantity of alcohol they consumed just while standing there impeding our view. The junior member of the debate team would polish off one beer, remove the can from its #48 coozie, slowly and methodically zip open the cooler, pull a beer out, and replace the former one. He'd sway a little, sometimes to the right, sometimes to the left or even backward, but he never fell or even tipped over. Junior repeated this move at least thirty times, and I never once saw him refill his cooler with more cans of beer. It was fascinating, like one of those little circus cars from which the clowns just keep coming and coming. Junior had a little magic circus cooler. Amazing. Now Senior, he was a different story. He did disappear with his Dale Earnhardt Sr. cooler every now and then, returning with it heavier than it looked when he left. And there the two of them would stand, sway, and drink, exchanging words occasionally, but mostly just methodically and deliberately drinking.

Something very interesting happened during the first caution of the race that day. Driver Denny Hamlin, making his return to racing after a four-week break that was the result of a fractured back he suffered in a California race about a month earlier, pulled off something that I thought was rather cool. Hamlin planned, before the race, to climb out through his car roof during the first caution—whenever that was—while driver and partner Brian Vickers slipped in through the window, a quick Houdini-like move in the pit that they had practiced repeatedly until they had the maneuver down to a minute or less. Hamlin hadn't planned the move for showmanship (though it was cool to watch him pop out of the roof while Vickers climbed in the window). He simply wanted to ease back into racing, see how his back reacted, and collect a few points while he was at it. Talladega's long track with its longer cautions was the perfect place to try the move. Vickers seamlessly reentered the race, and a mere eighteen laps later, The Big One happened.

The Big One, the big wreck, at Talladega this year was a thirteen-car smashup caused by driver Kurt Busch on lap forty-two. The wreck took out Busch, Kevin Harvick, Brian Vickers, Tony Stewart, Greg Biffle, Jeff Gordon, and several other drivers, some for the rest of the race, and some just until their cars could be hammered and taped back into racing condition. I was impressed with Busch's sportsmanship when he said first that he was all right, second that he caused the wreck. It's my observation that when something like that happens, even on a smaller scale, a diva-like football or basketball player will point first to someone else, anyone else, rather than himself. I must say, the crowd's reaction when those cars slammed into the wall, parts went flying, and damaged cars spun dangerously in front of others still racing, surprised me. They cheered in delight. They actually cheered, shouting things like, "How do ya like that?" and "There's more where that came from!" Granted, that was only the second really big wreck I'd ever seen live, but the reaction at Daytona was not quite like that. Of course, that wreck was much more serious too. I have to figure this out, why fans are so ravenous to see The Big One. I really don't believe they want to see anyone get killed or even seriously injured, but there is most definitely something about a wreck—the bigger the better—that gets those crowds jazzed and on their feet.

AS FOR THE REACTION OF MY NEW FRIENDS, I can only figure that Junior hallucinated that he heard the announcer blame the stupendous wreck on driver Kasey Kahne, who was involved but clearly not the cause. Junior, beer in hand (I never saw his hand without one the entire time we were there), rared back so far he nearly did a backbend and scraped his shaved head on the concrete, then shot forward and shouted at the screen in the infield, "That's bull----! That wasn't Kahne's fault! You ---- ----! That's bull----!" Then he turned around to address his woman with pretty much the same rhetoric. "Can you believe that crazy ------ ----? Ain't no way that was Kahne's fault!" The entire time he pontificated about the wreck, who caused the wreck, and what he thought should be done to said individual, he sprayed spittle and swayed dangerously back and forth, but never, not once, did he fall over. He twirled like a skinny pine tree in a tornado, dangerously close to losing his balance every time his weight shifted from dead center to the left or right (or backward, when he really wanted to emphasize a particular point). We all three cringed, just waiting for him to flip over the rail or fall straight back and crack his skull on the concrete, but he didn't. In fact, his feet never even lost contact with the concrete platform by the rail. It really was a fascinating thing to watch, like a demonstration straight out of Galileo's book *On Motion*. It was as if the guy's feet were superglued to the concrete.

Equally as fascinating was the fact that as my family and I sat there and ate our nerdy sandwiches and drank our nerdy water, Junior's family had sat down to eat too, even Senior. No one once jumped up to stabilize Junior or to admonish him to calm down or sit down. They just cracked open fresh beers and passed around a gallon bag of greasy shredded pork barbecue and another bag of turkey legs. They ate the barbecue out of the bag by the handfuls, like potato chips, and they just gnawed on the turkey legs until the bones were as clean as a whistle.

When the Alabama rains came that eventually caused a three-hour-plus rain delay, the whole clan vanished, the only telltale signs left behind were a few hundred beer cans and clean, white turkey leg bones.

I simply leaned over and pulled out the three rain ponchos we had responsibly purchased the day before. We slipped them on, and we began to wait. The first round of rain didn't last long and when it was over, out came the Air Titan™ track dryers. The trucks moved slowly and deliberately around the track, pass after pass, sounding like jet engines and blowing water off the track. They're louder than the cars and not nearly as exciting to watch, but they do get the job done. Unfortunately, just as the track was almost race-ready again, a torrential downpour like only Alabama can produce opened up over the Superspeedway and let loose with all its might. After a few minutes of that the red flag was waved, and that kicked off the three-and-a-half-hour rain delay.

When racing resumed around seven-ish Sunday evening, fans were ravenous for an exciting finish. They got it, with a win that took everyone, even the winning driver it seemed, by surprise. Driver David Ragan, who hails from Unadilla, Georgia, bamboozled Carl

Edwards with a final-lap pass that put the race away. Driver Matt Kenseth had led the pack for 142 laps of the race; Jimmie Johnson spent some time at the front of the pack, and pole-sitter Carl Edwards even looked as though he might pull a win out of his hat. But in the end, Ragan and his #34 Ford took the day. Talladega 2013 was only Ragan's second Sprint Cup win in seven seasons of chasing them, and both have been on restrictor plate tracks. The twenty-seven-year-old driver from Dooly County in south Georgia won his first race, the Coke Zero 400 at Daytona International Speedway, in 2011. What an amazing and unpredictable finish at the Superspeedway in Alabama. "It felt great to win at Talladega," Ragan told me. "It's a very popular track on the NASCAR schedule, and having a one and two finish with my teammate David Gilliland was extra special. Any win in the top three series of NASCAR is a big deal, but this one was for sure special." Now, when I hear a race fan talk about Talladega and grin that knowing grin, I'll know why. As the man said, "This is Talladega!"

I've had a while to think about this, and believe it or not, the closest comparison to which I can liken my experience at Talladega is the one I had many years ago at the Kentucky Derby. Both events are races, and at both events, drinking is an all-day affair. The attire of the fans was different for each race, and of course in Kentucky they were racing horses for two minutes, not cars for three-plus hours. This may sound like a strange thing to remember, but Kentucky Derby race fans are given real glasses from which to drink hoity-toity mint juleps and other delectable concoctions. At Talladega, there were a lot of aluminum cans and plastic cups. But no matter the vessel, it contained alcohol, and a lot of it. At the end of the day at Churchill Downs, I could hear glass breaking all over the place as inebriated race fans juggled and dropped their cock-tails. I may or may not have been one of them doing the juggling and dropping. But at Talladega, I just saw stray cans and cups on the ground after the race. You see, as race day stretches on and the mint juleps (or beers or tequila shots) add up, things get dropped. Very dangerous at Churchill Downs, but not so much at Talladega Superspeedway.

Chapter Eleven
Straight to the "Lady in Black"

The race at Darlington Raceway (the track also known as "The Lady in Black" and the track "Too Tough to Tame") came right on the heels of the Talladega race, the following weekend, in fact. I'm glad it did because in all honesty, my experience at Talladega had left me a little shaken. All the crazy stories I had heard were true, and those Alabama parties were a little over-the-top for this newborn recruit. I think I needed the Darlington experience.

This time, my husband and I hit the road for the Bojangles 500 with our friends Steve and Frances, who are tried-and-true NASCAR fans. In fact, Frances has tutored me all along this journey, feeding me facts and trivia, making sure I had the lingo down pat and filling me in on all the latest driver gossip. She's been a treasure. As long as those two have been NASCAR fans, they had never been to a race at Darlington Raceway. In a way, this was a first for us all. Bojangles Chicken, my husband's favorite fried chicken restaurant, was the sponsor of the 2013 race, so he was doubly excited at the prospect of indulging in their particular rendition of deep-fried fowl for an entire day. He may hail from Ohio, but he got the religion of eating fried chicken many years ago, once he and his family moved South. Southerners can take full credit for two things: we invented both fried chicken and sweet tea, the sweeter the better if you were to ask the older generation. I remember my mother at times asking me to make the tea for dinner when I was a young girl. Following her example, there were equal parts water and sugar in that recipe. It's unhealthy, and now I make the sugar-free version, but you can't really wash down fried chicken with anything other than a glass of sweet iced tea. Trust me on that.

Darlington, South Carolina, is a quaint little town about 290 miles from Atlanta. We wanted to get to the track in time to enjoy all the prerace festivities, and we had been invited to an "insiders" party by photographer and radio personality Doug Allen, a friend of mine who hosts the radio show "Spirit of Racing." He's another one who's been a helpful teacher to me, guiding introductions and inviting me on the air occasionally to interview guests and be interviewed on his show. Strangely enough, Doug is a transplanted Yankee and a graduate of the University of South Carolina. "I always considered NASCAR that redneck sport that people in the South watch instead of football on Sundays," he said, laughing as he remembered his early days in the South (Doug is still a South Carolina resident. I told you—they come to visit and end up staying). Working for a small newspaper in Elkin, North Carolina, at the time, he was asked years ago if he could fill in for another photographer and shoot some photos of a race one Sunday. "I thought, 'Why not?' I mean, I knew nothing about photography or NASCAR," he recalled. While Doug was strolling along pit road at that race, he got the opportunity to see some of the greats up close: Dale Earnhardt, Bill Elliott, Terry Labonte, and Rusty Wallace. He was intrigued, but still not impressed. And then, he saw Richard Petty. Even he, a Yankee without a clue about NASCAR, knew that Petty was the King. "I raised my camera to get the shot, and he actually stopped for me and posed. I couldn't believe it! I freaked, because the camera was out of film. To make matters worse, I had no idea how to reload it. Fortunately, a photographer standing near me did it for me, and Petty stood there waiting for me to take the picture. Richard Petty could have gone on his way, but he didn't, and I got the pictures. From that day on, I always had two TVs going in my house on Sundays—one for football, and one for NASCAR."

On the Saturday before Mother's Day, our little foursome hit the road at about 8:00 a.m. for a 7:00 p.m. race, just to be sure we could do all that we wanted to do before the green flag waved and the planes flew overhead. At Darlington, however, there was no military flyover. I guess the federal budget cuts are a fact of life now. The prerace invocation almost made up for the disappointing absence of the flyover, as First Baptist Church of Darlington pastor Mark Jones asked the Almighty, in thirty short seconds, for His favor on the military, on moms, on the drivers, and on the fans, even asking that God help the fans behave in a dignified manner at the race. Who else but a Southern Baptist preacher could tie up such a far-reaching request in a thirty-second prayer? I was really surprised by the difference between the setting for Darlington Raceway and that of Talladega Superspeedway. The Alabama track sits alone at the edge of Talladega National Forest, and it's the only structure of any consequence along that stretch of I-20. Darlington Raceway sits smack in the middle of a small town on land that was once used to produce cotton and peanuts. Built in 1949 by businessman Harold Brasington, it is a one-and-a-quarter-mile track that was originally intended to be a true oval shape, but landowner Sherman Ramsey didn't want his minnow pond disturbed. Brasington narrowed the track along Turns Three and Four to accommodate Ramsey's wishes and preserve the fishing hole, so Darlington Raceway is egg-shaped instead. It was also the first asphalt track for many of the drivers who drove in the 1950 inaugural race. Some of them ran out of tires well before the race was over, since asphalt chews through tires much faster than dirt. I was again reminded of the resourcefulness of NASCAR teams when I read that some of those drivers in that first Darlington race resorted to buying tires from fans in the packed-out grandstands, just to be able to finish. That's where the track's nickname "Too Tough to Tame" came from. That first race, called the Southern 500 and won by Californian Johnny Mantz, took six hours to complete. The average speed during those six hours was a scorching 76 mph.

I could sense a difference between the two race personalities of Talladega and Darlington, even moreso than I could between

At today's NASCAR races, teams come prepared to chew through a lot of tires.

Daytona and Talladega. Maybe the bright, warm, sunny day in Darlington had something to do with that. Still, the fans were more, shall we say, laid back at Darlington. They displayed Southern-style manners, and I love that. I saw more of the graciousness I experienced at Daytona, with people sharing food and drink between campsites or with passersby, laughing, and self-policing. What I mean by "self-policing" is that, if someone did get a little rowdy or otherwise out of hand, the people around him would get the situation in check before it escalated; in other words, no harm, no foul. In a nutshell, I felt very safe and in my comfort zone at that race in Darlington. Sure, there was drinking and partying (the tailgating was absolutely fabulous), but it was all very civilized from what I saw.

Before we attended the "insider party," we walked up and down the aisles of merchandise booths and food vendors. I bought another Clint Bowyer shirt and a pair of sunglasses with black-and-white-checkered frames. They're positively hideous, and I love them. I ate some kind of meat on a stick, which was delicious. I won a free T-shirt and collected more free race day swag—signs and pens and sunglasses and hand sanitizer. A huge fan of cool free stuff,

Campers at Darlington race.

I just love that part of any race. The Budweiser Clydesdale horses
and the famous spotted dog Travis the Dalmatian were there that
day too. Those majestic equestrian beauties are even more impressive up close than they are in all of those wonderful, tear-jerking
commercials you see on television during football season. Once we
had shopped and done our own mini-version of parking lot tailgating (sandwiches, pasta salad, fruit, cookies, and generous chilly
glasses of Frances's world-famous sangria), we headed over to meet
Doug for another get-together near the track. That one event, in
my opinion and up to that point, helped me define the quintessential NASCAR fan. The people in attendance were longtime fans
who plan to attend the Darlington race, no matter what, every year.
At that cookout, I got the chance to meet and talk with Ted and
Cheryl, who had traveled from Canada to see the race. Every year,
they meet and camp with another couple from Pennsylvania who
also make the annual pilgrimage to Darlington. Their friendship
formed eighteen years ago, when they were tailgating neighbors
at that same track. I met another couple, Glenda and Malcolm,
who are Tony Stewart fans through and through. For her birth-
day this year, Malcolm gave Glenda a customized golf cart with

the word "Smoke" painted across the front ("Smoke" is Stewart's nickname). Malcolm's elderly father James—a NASCAR fan from way back—was also there that afternoon, and he shared with me some great stories. "I remember going to the races in sixty-eight and sixty-nine, and when they were over we'd walk around picking up Pepsi and Coke bottles to pay for the gas to get home," James said, laughing. He remembers a race in which Bill Elliott's car got stuck in the guardrail surrounding the track, well before the days of cable-reinforced catch fences. He remembers building scaffolding with his buddies, then climbing to the top with his AM/FM radio to be able to see and hear the race from outside the track. NASCAR has been an important part of James's life for most of it. I've learned that that's true of a lot of fans. They love the sport with a passion, and they've passed that love on to their children. NASCAR is counting on a new generation of fans joining the ranks, as well. Fans, drivers, and everyone else I've asked have some very definite ideas as to how that will be accomplished.

I also got to spend some time at that party talking with Don, a devout fan who still carries a Darlington race day ticket for his wife every year since her passing five years ago. He showed me his prized possession, a lovingly restored 1969 Mercury Cyclone called a "Cale Yarborough Special," one of the few that were built with a black leather interior instead of the characteristic red leather covering. Apparently, in 1969 Mercury simply "ran out" of red leather for the seats of those cars. Don proudly showed me the paperwork trail that led to his owning that car, along with "before" and "after" photos of the showpiece. I think, at that one small gathering, I got the absolute best snapshot of what being a NASCAR fan is all about.

When 7:00 rolled around that Saturday, we were already seated and ready for the race. Here's another difference that made that race a much better experience for me: we sat much higher up in the stands, and we made sure we were seated right across from pit road. What a difference it makes to be able to see those pit crews stand on the wall at their driver's pit, crouched and ready to do their jobs. When the driver "pits" the car, they practically jump in

Sunoco fuel tanks, used to fuel up cars before races and in the pits. The design shaved seconds off pit times.

front of it before it even stops, jack up the car, change four tires, fill the gas tank, and do dozens of other things that I couldn't catch in a dazzling twelve seconds or so. Instead of washing the windshield as they did many years ago, one crew member pulls off a thin plastic coating, clearing the glass. As they run to jump back over the wall, the car slams back down on all fours with tires spinning, and screams out onto the track to get back into the fray. Every time I see it, I am fascinated. I found myself timing the crews with my watch that night, eyes wide and breath accelerated. I really think watching the pit crews is every bit as breathtaking as watching the cars fly around the track. Just as Ed Clark had said to me months ago, there's an added element of excitement at a night race. The lights of the track wink and reflect off the cars, and at this particular race, it seemed to me that the pace was much faster. Of course it wasn't, but remember that the Darlington track is a mile or so shorter than both the Talladega and Daytona tracks. It just seemed to me that the cars raced even faster on a shorter track.

For most of the race, Kyle Busch and Jimmie Johnson led the pack. Fans surrounding us in the stands went wild as the number-one spot on the leader board changed back and forth between the two, the laps mounting up. Very few caution flags were thrown (another contrast to the other superspeedway races), and by the time the race was four hundred or so miles old, even I was convinced Johnson or Busch would win, hands down. Bowyer, always my favorite, stayed in the top ten on the board for most of the race, but he was penalized for an infraction on pit road (something to do with the gas tank, but that's all I could hear over the crowd). He did battle back to finish eleventh. With about thirty laps to go, Busch dropped back because of a cut tire, and shortly after that, Matt Kenseth came out of nowhere to overtake everyone and win the race. I was glad for him, because he led at Talladega for 142 laps and still didn't win. Driver Denny Hamlin, who had been nursing a fractured back for a month or so (remember, he and Vickers pulled the Houdini driver switch at Talladega), came from behind to take second place. I noticed at Darlington that the drivers did not race three and four wide, as they did at the longer tracks. This race looked more like a long line of cars pretty evenly spread around the mile and a quarter track, at least until a driver decided to make a move. It made me less nervous, as those tight bunches of cars are just an accident waiting to happen, in my newly formed opinion. The last twenty laps of that race, as I'm finding is true of many, were the most thrilling. When the race was over, we made our way out of the stands, down the steep steps, and back to the car. I'll confess, Frances and I rented a rickshaw of sorts to take us out to our car in the parking lot. We were exhausted, but we were happy too. The only drawback to this entire experience was that we did not rent hotel rooms, but opted to drive back to Atlanta that same night. Two hundred ninety miles looks a lot shorter on a map than it is in real life, especially at the end of a long, sweaty day. Our adventure turned out to be almost a twenty-four-hour one, so if I'd change anything about going to the Mother's Day Weekend race in Darlington, that would be it. Next time we'll camp, or better yet, we'll book a room in a nearby hotel. Camping is something we hadn't done just yet.

The one race I wasn't able to see live in the 2013 season came later that month, the Memorial Day race in Charlotte, North

Carolina. Named the Coca-Cola 600, this race has a reputation for being an exciting, patriotic celebration of both stock car racing and the United States military. This race was one that I really wanted to see (about midseason, I found that I experienced a sort of withdrawal if I went a couple of weekends without being at a race), but a scheduling conflict prevented me from going. This particular race is the longest NASCAR-sanctioned race of the entire season, stretching out for a brutal six hundred miles. It starts in the daytime and ends in the evening, and there is pageantry and patriotism galore. I love military pomp and circumstance, the red, white, and blue, the bands and the fireworks—all just more reasons NASCAR and I have turned out to be a pretty good fit. The 2013 Memorial Day race was billed as the World's Biggest Memorial Day Party, with a prerace pizza party for ten thousand members of the armed forces and their families. In fact, John Schnatter, founder and CEO of Louisville, Kentucky–based Papa John's Pizza, catered the "Welcome Home" event and even helped serve pizza to members of the military at the race. As further demonstration of his organization's support of the troops, Schnatter raced his beloved Z28 Camaro against Charlotte Motor Speedway president and general manager Marcus Smith's Ford Mustang, for the tongue-in-cheek "Golden Hubcap" Award. The friendly competition resulted in raising $50,000 for the USO. During the race festivities, the focus was entirely on honoring and thanking military personnel for their service. Even many of the drivers' cars sported a red, white, and blue theme. Military families got the chance to mingle with former Lt. Colonel Oliver North, drivers Greg Biffle, Kurt Busch, and a few others, and HLN host Robin Meade, who emceed the party and sang the national anthem before the race. As an added treat Willie Robertson, from the hit reality television show *Duck Dynasty*, was on hand that day to greet and thank service members. Vintage military planes performed the privately funded flyover of the packed stadium, and a bald eagle circled the infield before the race. Six hundred troops marched through the stands, and buses of military personnel circled the track before the race. NASCAR loves the military, and the military, it seems, loves NASCAR. Troops and their families, with fans who made a special trip to Charlotte just to say thank you, filled the grandstands that day, and I would have loved to have been there myself. Of course, I watched the race from start to finish on television, but I'm sure that experience didn't hold a candle to seeing the whole thing live. By all accounts of the people who described the experience to me, it was thrilling.

I also really wanted to see the track itself, which has its own fascinating, checkered history. Legend has it that in 1959, track owner Bruton Smith and a gentleman named Curtis Turner (a former moonshine runner later dubbed, "The Babe Ruth of Stock Car Racing") joined forces to build a one-and-a-half-mile track with forty-five thousand seats in the Charlotte area. They expected the cost to be right around $750,000. Shortly after Smith and Turner joined forces, they signed a contract with NASCAR to run the six-hundred-mile race on Memorial Day. Here's where the story gets interesting, as I've learned that many of these tales of early NASCAR escapades often did. Once the construction crew began working on the project, they found that a thick layer of granite ran just under the soil, and the cost of blasting the stubborn rock out of the ground on the first turn alone cost a whopping $70,000 just in dynamite, as well as precious additional time. Figuring in the higher costs uncovered by the hidden granite, the price tag for the entire project soared from a mere $750,000 to about $2,000,000. An unexpected snowstorm that same year (1960) delayed the pouring of concrete to complete the track and pushed the project back six weeks, leaving Turner with no other option than to plead with NASCAR for a six-week postponement of the inaugural race. The cherry was placed on top of this disastrous cake when the paving contractor threatened to walk off the job because of his claim of a lack of payment. The story goes that Turner, no doubt already at his wits' end, enlisted the help of a friend, a shotgun, and a revolver to convince the contractor to stay on the job and complete the unfinished work on the backstretch. As a result of all the delays, the bad weather, and a reluctant contractor, the first "Memorial Day" race at Charlotte Motor Speedway was actually held on June 19, 1960. I assume that the paving contractor ultimately got paid in that deal. I found several accounts of this piece of Charlotte track history in my research, and they all told pretty much the same story.

Years later, when lights were installed around the track, fans petitioned NASCAR to move the start time of the race to a time later in the day. The South, even as far north as North Carolina, serves up heat and humidity in the summer like no place I've ever been. With even the last part of a race taking place at night, fans could enjoy the cool relief that only nighttime brings. I share all of this history to set the stage for a race that rarely disappoints with respect to entertainment and excitement. The Charlotte Motor Speedway has seen fiery crashes, wrecks that have cost drivers, literally, life and limb, a fuel crisis in the 1970s, wild lead changes in many races, and controversial finishes. The 2013 Memorial Day race, then, was no exception.

On lap 121 or so of the four-hundred-lap race this year, a thick nylon cable used to suspend a Fox Sports camera over the racing action and pit road, snapped and fell onto the track. The caution flag was thrown, then the red, stopping all racing action. Ten spectators were injured by the broken cable, with three of them being carted off to a local hospital. One account of the event, chronicled in Charlotte's NPR News source, quoted a fan who witnessed the whole scene as saying that a car actually ran over the nylon cable, causing it to wrap around one of the axles. The rope stretched, then snapped back into the crowd, and that's what caused the injuries. Cars were also damaged as a result of the cable snapping. The red flag was out for nearly a half hour, giving teams a chance to repair their cars and get them back out onto the track in better shape once racing resumed. The incident was eerily reminiscent, said one fan, of the May 2000 mishap in which more than one hundred fans were injured outside the speedway when a section of the fan walkway fell twenty-five feet onto the highway below.

A second red flag in the 2013 race was thrown when Dale Earnhardt Jr. blew an engine, causing #16 driver Greg Biffle, along with a couple of other drivers, to crash into the wall. Three laps before that smashup, Kyle Busch's engine failed. Finally, on lap 325, drivers Aric Almirola (#43), Mark Martin (#55), and Jeff Gordon

(#24), racing three wide and each refusing to back off, mixed it up and caused a domino-like pileup that involved several drivers behind them. Kevin Harvick (#33) ultimately prevailed in that exhausting, exciting battle in Charlotte, winning the Memorial Day race at Charlotte Motor Speedway for the fourth time in his career.

Yes, next year, I am going to make it a point to be at the Memorial Day race in Charlotte. Who wouldn't find all that action to be exciting and riveting? As so many drivers who understand the lure of speed and power have told me, experiencing a race live and in person is exciting. It's addictive. I agree; I believe that new fans will be brought into the fold simply by being introduced to the rush of live racing, to that that intoxicating sound of unbridled horsepower roaring to life. I'm a living testament to that, and I don't even need a wreck, a feud, or a broken camera cable to get my pulse racing.

The notion of the next generation of fans and how they will be cultivated is an important one to the powers-that-be in NASCAR, and it matters just as much to the fans themselves. In the past, fans have loved NASCAR simply because their parents and grandparents loved the sport. I think it's even safe to say that NASCAR is a beloved tradition in many families, just like having Sunday dinner together or serving turkey on Thanksgiving are family traditions. But what about the future of the sport; where will the next generation of fans come from? As with any sport, fans age out, and the life of the sport depends on new ones being brought in. There are numerous opinions as to how NASCAR will attract the next generation of fans, many of whom have never laid eyes on a NASCAR track, either in person or on television.

Chapter Twelve
Who Will the New NASCAR Fan Be?

A common question that arose no matter who I talked to throughout this adventure was, "How will NASCAR attract a whole new future wave of fans?" This challenge is not unique to NASCAR; any industry depends on continually bringing in new fans, customers or clients. The unique challenge is how, exactly, to best do that. This new generation of young people is an entirely different animal than any sport or industry has ever approached before. They will not be reached in conventional ways, that's a given. They are the first generation that's always had cell phones and computers as part of their everyday lives. This generation of consumers-to-be, often referred to as "Generation Z," learned the alphabet watching Kermit and Elmo singing and dancing on a computer screen. They perform research for school term papers by searching for information on the "Wild Wild Web" (a term for the Internet that's growing in popularity and accuracy); the solid, calming security of reading and turning the pages of an Encyclopedia Britannica or World Book is foreign to this generation. Those vast treasure troves of information, so dear to me in my own childhood, now contain outdated material by the time they've been printed. This wave of young people takes their very own cell phones to school with them, and they've done that since about first or second grade. They play video games instead of actual physical games. They often prefer texting to conversation, and they can do it at the speed of light, often to multiple friends at the same time. In fact, a group of them can be sitting in the same room for hours on end in complete silence; they're all texting each other as well as friends in other rooms, other states, or even other countries. I know what I'm talking about. We have four children ourselves, and they're now all in their twenties. When they were teens, we'd often have summer parties, our pool packed edge to edge with dozens of goofy, roughhousing teenagers. I can recall many times though, walking out to see who was doing what, making sure they had some kind of food and drink—nonchalantly spying, in other words. They'd all be sitting on chaise lounge chairs, sunning and texting, no one saying a word. It still seems terribly rude to me (my mother always told us that it was rude to whisper in the presence of someone else and to me, texting is electronic whispering), but they don't see anything wrong with it. Clint Bowyer probably described today's young people in the simplest terms by telling me, "They're fast, and they're used to being entertained all the time." That's a good point, and it's one that presents quite a conundrum for businesses vying for their attention and ultimately, their dollars. "It doesn't matter whether it's NASCAR or anything else, we have to change with the times. We have to change how we reach this younger demographic but most importantly, we have to entertain," Bowyer explained.

Fans, too, have had plenty to say to me on the subject of bringing in the next generation of dedicated NASCAR enthusiasts. Longtime fans, probably those who have the purest appreciation of old-school racing, are wary of changes they already see being made in the sport. For instance, many of them think that the advanced safety measures on the tracks and in the cars are watering down the racing experience. I really did find that idea to be strange. As much as diehard fans love their drivers, a lot of those who have been around for a long time want to see racing "as it was meant to be," as J. R. in Talladega told me. "It's not racing if you take all the danger out. That's what makes it exciting to be at a race, knowing that at any minute, there could be a wreck that takes out a few cars."

Clint Bowyer, suited up before a race.

The bigger the smashup, it appears, the wilder the excitement. Sparks flying, flames licking out from underneath a car, quarter panels and bumpers crumpled or dragging—they're all crowd-pleasers. I guess I'm just too inexperienced still, at this point, to really appreciate the value of a smoking, flaming wreck. Perhaps that will come with time.

I have been told time and again throughout this journey that many longtime fans feel that NASCAR is slowly turning its back on them, the very people who built it into the powerhouse that it is today, just to chase after the numbers and money of the "yuppie" demographic. As I explained earlier, I know that the term "yuppie" is a relic from the 1980s, but it still lives on the lips of resentful fans who feel that their support and dollars are not enough for the "higher-ups" (another term I hear a lot from fans) of NASCAR. Popular NASCAR radio personality Dallas McCade had something to say about that matter, as a longtime fan and broadcasting darling of the sport: "I do think NASCAR is trying to tap into the yuppie fan base, and why not? They have the good ol' boys; let's get the yuppie money too. I think there's room for all of us." Wise words, and from a woman who's loved NASCAR practically all her life.

No matter what phrases are used when these concerns are aired, some fans feel that NASCAR is homogenizing the sport, smoothing off the edges, to make it more palatable to a younger, more white-collar fan base with fatter wallets (or newer plastic with higher limits). I must say here again that no one, not one single person affiliated with NASCAR and with whom I've spoken, has voiced the opinion that its oldest fans don't matter. From the top down, it seems to me that NASCAR values its longtime fans and understands that without them, the sport could not survive.

One thing on which everyone agrees is that excitement is an adrenaline rush, and adrenaline can be addictive. The fans love it. I have to admit that I love it, and I never thought I'd say those words about racing. In fact, not only have I not been a fan of NASCAR all my life, but prior to this journey, I was a reluctant observer of the sport. While I committed to learn every last thing about it that I could, I must say that I was not as excited about it as I might have been about, say, the opening of a new outlet mall or traveling to exotic destinations. Even so, the cars, the excitement of going to the races, and even the accessibility of the people have all come together to make me a true NASCAR fan. I think Johanna Long may have hit the nail on the head when she offered me her opinion of the best way to go about bringing new fans to NASCAR. "Just get them to a race, one race. That's all it takes. When you hear the cars, smell the smells, and get caught up in the competition, that's what it's all about. One race, and people are hooked." And she may be right. The speed and the adrenaline, and those oh-so-powerful and incredibly sexy cars, are what made a believer out of me.

I had the opportunity to interview a sixteen-year-old driver named Mason Massey. Oh, he's too young to drive NASCAR—yet—but make no mistake, NASCAR is watching him. Mason is a good-looking kid from Douglasville, Georgia. He's home-schooled, and that allows him to travel and race when he needs to. Right now, he races super-late-model cars, but he has his eye on NASCAR. He thinks that fans need to have more opportunities to meet the drivers in NASCAR, and that might bring even more fans into the fold. He may be right, but my money's on young kids like him to be the key to attracting younger fans. He started racing at age five (quarter midget cars) with the support of his entire family. He loves racing, he loves his car and understands how much money it costs to repair it, and he loves his family. This young man is not only a fearless driver; he's a role model for people his age and younger. If you follow NASCAR at all, remember his name.

When I was at the Darlington race, I had the opportunity to talk to two young brothers, Joshua and Jason. One is in his late twenties and the other in his early thirties. While they munched on chicken wings and barbecued ribs that smelled simply out of this world, I asked them how they thought NASCAR will bring in younger fans. Both of them answered without hesitation, "They'll have to use technology." Both of these guys inherited their love of NASCAR from family elders—dad, grandpas, aunts, and uncles. That's how a good many NASCAR fans came to love the sport; they were taught to love it. McCade feels strongly that that's exactly how the next generation of fans will come to NASCAR, and she shared with me her own very definite ideas on the subject. "NASCAR fans are true fans of the sport, not fair-weather fans. Young people who grew up with their families being fans will be the ones to keep the sport alive." McCade knows what she's talking about, having been a NASCAR fan herself since she was eleven years old. "I went to my first race at Talladega with my Girl Scout troop and have been a fan ever since. I have been in radio since the age of fourteen and always incorporated NASCAR into my air time, because it's been such a hot sport in the South." *Now that's a fan*, I thought as the vivacious blonde with a stunner of a smile shared with me her passion for the sport. I also thought that she must have belonged to a pretty darned cool Girl Scout troop for a Talladega race to be among their list of approved field trips. Even so, Joshua and Jason seemed to have their own opinions of how a new young fan might be brought on board. Let's go back to their suggestion that technology will have to be a part of the bait; in order to get the attention of most young people these days, you'd better have a phone, a computer, or a game in your hands to help you out. Enter something called iRacing. "It's a video game, but it's not really a game. It's a racing simulator," said Jay Guy, the pit crew chief for David Ragan's team. According to him (and he ought to know), iRacing is a great racing simulator that looks very much like a game. What does today's younger generation love more than games, especially those that start with "i-"?

Guy went to a NASCAR race in Dover with his family for the first time when he was seven years old. He and his dad met driver Dave Marcis. The guys became friends, and eventually Guy and his dad began working on Marcis's team off and on, then summers, all the while learning the business of cars from the ground up. "It's hard for me to understand what it's like to not be around [NASCAR] all the time. It's all I've ever done. I've never had a real job that wasn't associated with NASCAR," said Guy. However, when asked about iRacing, he said that the driving simulator is a good way for a young person to familiarize himself with tracks, speed, and maneuvering. Many young fans and even aspiring drivers have been introduced to racing through iRacing.

What else is going to make NASCAR an attractive new experience for a new generation of future fans? "Going to a race has to be affordable, number one," said Joshua. "College students never have a lot of money. And they're going to have to bring in social media. There's no other way to reach young people these days."

That's yet another good point, and I must say that I think the marketing folks at NASCAR have already plugged into the power of social media, from e-mails to Facebook to Twitter. When I decided to learn all I could about NASCAR, I signed up to receive e-mails from official websites, media sites, any source of information I could think of. I "liked" the NASCAR Facebook page, and

I followed every driver whose name I knew on Twitter, as well as some fans. Social media has played a big part in my education in stock car racing, as well as my knowledge of upcoming events, ticket specials, and race weekend entertainment.

Driver David Ragan, who walked away a winner from that wild race at Talladega this year, also offered his thoughts on the ever-present challenge of hooking and reeling in new NASCAR enthusiasts. "I think in order to bring in the next generation of fans, we have to keep the action at a high on the track, and we need to have good personalities off the track. I think fans will always love to hate certain drivers, pull for their favorite car manufacturer, and continue to come to the track as long as the experience is fun at a decent price. We have to stay loyal to certain traditions we've had over the years, but the sport needs to grow from a technology standpoint, just as the world we live in grows." Ragan also threw this idea out there: "Some of the races are too long and need to be shortened." *Interesting.* I have thought the same thing about the difficulty of keeping a young person's attention through the three or four hours (sometimes longer, with delays) it takes to run a race, especially a six-hundred-mile race like the Memorial Day slug-fest in Charlotte. With the instant gratification of faster-than-sound communication, some young people's love relationships don't last three or four hours. Is it possible for them to commit that much time to a pastime, however exciting it may be? Maybe the answer to the question of bringing in new fans involves offering a few shorter races throughout the season. It might be something to think about.

Ed Clark addressed this very question with me during a conversation we had earlier in the season. A Legends car driver himself, he is passionate about racing in general, NASCAR in particular. "We are aware that attendance has dropped off a bit in recent years. I believe that is due, in part, to the economy. But I also know that we have to attract new fans. We know that our longtime fans are critically important to NASCAR, but it's also very evident that we have to bring in fans from a more diverse population, and a younger population. Otherwise, the fans we have and enjoy now will essentially just age out." Brian France, in a May

25, 2013, media conference, addressed the issue of slightly flagging attendance, and he used Clark and Atlanta Motor Speedway as an example of a success: "Ed Clark is running quite a bit up, as a matter of fact, in Atlanta as an example," France stated. With drag racing also offered at AMS, along with other events such as a June Zombie Run 5K and driving schools (I had my first ride along experience there, remember?), the track has become a destination for Georgians, including suburban Atlantans, throughout the year, not just on race weekends. I attended a drag race there in June, and I was surprised to see the number of fans, many of them families with young children, who turned out for the show. The event was reasonably priced ($8.00 per ticket, which is less expensive than a movie), the music was great, and the competition was interesting. There were motorcycles, ATVs, cars, and trucks racing on the strip, and there was even a kids' battery-operated car competition about halfway through the evening. Atlanta Motor Speedway is a familiar place now, and it's therefore the place where many fans will see their first-ever NASCAR race. It's a story I've heard over and over as I talk to fans.

I have read and heard about monumental changes and shifts that have characterized the sport since its birth in 1947, yet today it still reigns supreme among professional motorsports. I can't help believing that a tidal wave of new fans, both younger and international, will discover the speed, the power, the "show" of NASCAR, and fall in love. Ambrose shared his thoughts with me on the development of an international fan base. As an Australian driver, his insights are keen, and they make sense. "NASCAR is running some support series in Europe now, and I think that's been good. We race in Canada and have raced in Mexico too, with a lot of fanfare. The television in other countries is getting better; a lot of fans worldwide can watch the NASCAR races live. I think we're seeing more and more fans of the sport globally. You have myself from Australia and Juan Pablo Montoya from Columbia. In the Truck Series and Nationwide Series we see other drivers from Australia or Brazil. I think we are really starting to see a more diverse driver base. Look at how many drivers are today from the

Southeast—not many. We have drivers from all across the United States. Now, we're starting to see drivers from all around the world. And drivers from different forms of racing such as V8, IndyCar, Formula One are coming to NASCAR to race. That's another way we are growing fans." Until Ambrose pointed out that so many of today's new drivers hail from outside the Southeastern United States, that realization hadn't occurred to me. The passion and dream of driving in NASCAR has spread to young people throughout the country, and throughout the world. As is true of many sweeping changes, the growth of NASCAR will take root in young people, many of whom will drive their first go-kart on a local track in their hometown this year.

One thing is certain. If NASCAR can make a fan of me, a fifty-something technologically challenged writer who used to hate loud noises, crowds, and danger, they can make a fan out of just about anybody. As Johanna Long said, "Just get 'em to a race." The power, the excitement, the entire experience of seeing a race in person will be enough to bring many new fans into the stands. In my opinion, the fact that NASCAR is taking the lead with initiatives that effect true global change will attract still more race enthusiasts. So let's assume that the NASCAR fan base will likely change, that the overall fan base may look a bit different but be no less enthusiastic, in the near future. What about NASCAR itself, though? Will there be any changes in racing?

Chapter Thirteen
NASCAR's Changing Face

A common thread that's run throughout my out-of-the-box NASCAR adventure this year has been a concern on the part of longtime fans that NASCAR is changing. Change is never easy. I don't even like changing my clocks twice a year to accommodate daylight saving time, because doing so represents change. So I get it when fans of this sport so deeply rooted in tradition and rich history express their displeasure whenever they see their beloved NASCAR changing. I think that displeasure is rooted in fear, not that any of the tough, stuck-in-their-ways guys with whom I've spoken would ever admit that. I think they fear that, in order to entice new, "different" fans to the sport, they will neglect the old fans and change NASCAR so much that it won't be recognizable to a diehard. As a brand-new fan but card-carrying fellow change-hater myself, I am going to say that the recent and even not-so-recent changes I've heard and read about in NASCAR are not just good, they're great. And they're necessary.

First, I have to say that driver and now team owner, author, and network broadcaster Michael Waltrip really put things in perspective for me when we talked about how the sport has changed just on his watch. Born in Owensboro, Kentucky, in 1963, Michael is racing-great Darrell Waltrip's younger brother. As a driver, he's no slouch himself. I suppose you could say he's seen a lifetime, and then some, packed with changes in the sport his family loves so passionately. "When I started racing, the races weren't available to fans through any media other than radio. Now, races are covered live on TV. They're even covered on the Internet. When I was racing, we'd have twenty-five thousand to thirty-five thousand people in the stands watching. Now, there are more than a hundred thousand fans at almost every race," Waltrip said. The racing legend also added that the schedule is more grueling now than ever, with more races per season now than when he was racing. Most important, Waltrip told me that the sport itself has changed because the skill level of the drivers has been amped up considerably. "The competition is tougher now than ever before. More people can win now than ever before, because there are that many really good drivers out there today." Waltrip added something else that I hadn't thought of before. "They're also reinventing the stadiums, making events more fun for fans, more interactive." I don't know what the stadiums were like before 2013, as I had never set foot in one, but I can say that the ones I've visited on my journey this year have been big and clean, offering fans anything they want, from barbecue to pizza, burgers to catfish (and more elegant fare in the suites), beer, wine, and cocktails. Fans can now request seats in the family section of many stadiums (a bit of knowledge that would have come in handy for me in Talladega). This change has been made for obvious reasons, as it is not uncommon to be surrounded by smoking, drinking, and a lot of bead-collecting by women who are not shy about complying with the "show us yer boobs" requests, which are made pretty often before, during, and after a race. Some women do not even wait to be asked; they just jump the gun and share with other fans what God or a generous cosmetic surgeon gave them. Family seating sections were actually a stroke of genius on someone's part. As a matter of fact, later in the 2013 season, Atlanta Motor Speedway made the announcement that smoking in the stands will no longer be permitted, except in designated areas. If NASCAR wants to continue to be considered a family-friendly sport, they're going to have to rein in some of the "friendly" and focus more on the "family." The special seating designated just for families was a good start.

Big screens in the infield and leaderboards that keep up with every driver also make the live race experience a fun one. Tracks are often so big that fans can't see all the action from where they're sitting. Even on the shorter tracks, it's nice to be able to see the skilled maneuvers and, yes, slow-motion replays of crashes.

These changes enhance the fan experience just on race day. Throughout the year, the stadiums offer fun and affordable entertainment to surrounding communities. For instance, at Atlanta Motor Speedway, summer Friday nights mean exciting drag races. In June, there's a 5K Zombie Run. Participants can be zombies (complete with makeup) or humans as they race around the track, which will be transformed into a Hollywood movie set. After the race, there's a "Quarantine Party," with a live DJ, music, beer, and food. Participation is projected to be great based on the number of runner/walker packets already picked up, and a whole new demographic (young people who exercise and are fans of zombies, vampires, and the like) will be introduced to Atlanta Motor Speedway and perhaps to racing and NASCAR. Other tracks offer the same types of community entertainment events, a smart move for a couple of reasons. First, there are many doors through which new NASCAR fans can walk. If one of those doors is a Zombie 5K or a first-ever experience at a drag race, then everybody wins. Two, events offered between races and sometimes year-round help tracks generate more activity and therefore more income.

I have actually had some old-time fans tell me that the downfall of NASCAR began when television started covering the races. Again, purists and change-haters cling to their ways (I know, I'm one of them), but honestly, television can only have broadened the scope and reach, and therefore the popularity, of NASCAR. Rising popularity of any sport attracts more athletes and better athletes. Featuring better athletes means more exciting competition and that, of course, results in stronger television ratings. We all know what great ratings lead to: money. Change can be good, and it can coexist with tradition.

I have always believed that the wider an organization's influence, the heavier the burden of responsibility on that organization to use that influence for good. NASCAR's influence, with its viewership, tremendous fan base, and across-the-board command of network and sports channels during race season weekends, is far-reaching. The drivers, as idolized public figures, carry great influence. Whatever message NASCAR sends out into the universe scatters like dandelion seeds and settles on millions of people worldwide. Let's face it, for many years what NASCAR sent out into the universe— intentionally or unintentionally—was an image of beer-drinking, cigarette smoking, tobacco-chewing, reckless, car-smashing "rednecks." The whole bootlegging thing just added to the perception that NASCAR was a spectacle for just such people, not a mainstream sport that requires skill, courage, smarts, and training. I'm not judging. I'm simply repeating what's already in the history books with respect to NASCAR's image in the past. Some of NASCAR's most devout fans will tell you the same thing about the sport's reputation and, in fact, that's what many of them like most about NASCAR. They like that wild, somewhat politically incorrect, borderline outlaw mystique that surrounded the sport back in the old days. They do not like when changes come along that threaten to dilute that image.

I came across an article in my research, an editorial piece really, written in 2005 by a man named Geoffrey Norman. This particular article appeared in the *National Review Online*, for which Norman was a sports writer. In his musings, Norman marveled at the fact that the *New York Times* covered the Daytona 500 race that year. He marveled that Fox's coverage of the race garnered higher ratings than that year's NBA All-Star game. He wrote, "What was once the pastime of moonshine-running rednecks has gone mainstream. NASCAR is even listed on the New York Stock Exchange." In this same piece, Norman compared the NASCAR of yesterday— epitomized by Junior Johnson and his famous bootlegging and moonshining escapades—to today's NASCAR, the face of which some say belongs to none other than the driver of the #24 car, Jeff Gordon. With the countenance, voice, and presence that could just as easily pass as those of a stockbroker and not a race car driver, Gordon "can flat out drive a car," according to Norman. I myself have read quite a lot about Jeff Gordon. I've watched his postrace interviews very closely. He's not scruffy. I have never seen him spit tobacco and wipe his mouth with the back of his hand during an interview, and he speaks with articulation and without profanity. Proudly open about his spiritual beliefs, Gordon is a Christian. In 1999, he established the Jeff Gordon Children's Foundation to help support children facing life-threatening and chronic illnesses. In 2007, he teamed with other world-class athletes to found Athletes for Hope an organization that connects professional athletes with charitable causes and encourages grassroots volunteerism. For heaven's sake, the man is from California. Married and the father of two children, Gordon is, quite literally, squeaky clean. Norman commented with regard to the driver's image, "It is the image NASCAR wants these days, but it ain't the same." And therein lies the rub: "It ain't the same." In fact, there are hardcore NASCAR fans who will likely never accept Gordon as being authentic, traditional NASCAR, but that doesn't change the fact that he's a stupendously talented driver, competing in what some say is the toughest motorsport in the world. When all the hype and shenanigans and personalities are pushed aside, isn't that what NASCAR really is? NASCAR is fast drivers driving fast cars, often to the exhaustion of both. The fastest, most clever drivers backed up by the most talented teams, and sometimes a little luck, win races. Gordon has the most wins in NASCAR's "modern era" (from 1972 to the present), and he became the first driver ever to top one hundred million dollars in career winnings. That, my friends, buys an awful lot of authenticity in my book. He doesn't have a Southern accent or the telltale circle on the back pocket of his jeans that a can of Skoal or Copenhagen leaves (I looked for that too), but the man can drive a car with the best of them.

From what I've gathered through this experience of immersing myself in NASCAR, what unites fans of yesterday and today, from the West Coast to the East, is a love affair with horsepower. It's an affinity for cars and the skill of those who push and coax those cars to their limits to win races. It's the roar of forty-three of those cars, powered by both brilliance and muscle, battling during a race, one car inflicting injury on another, then that same car licking its wounds in the pit when injured by yet another. When you look at it like that, surely both old and new can all get along because above all else, NASCAR fans love the cars and the competition. Besides, some changes will make the sport, the athletes, and the fans better.

During the Toyota Owners 400, the April 2013 race at Richmond International Raceway, fans were recognized and rewarded for (of all things) responsibility. Miller Lite beer and the Techniques for Effective Alcohol Management (TEAM) Coalition joined forces that weekend to promote responsible drinking and traffic safety at the Raceway. Before that race in April, 492 NASCAR fans pledged to be the designated drivers for their groups. Free souvenir photos and other prizes, as well as a chance to be chosen as the Designated Driver of the Race, were their reward for making the socially responsible choice not to drink alcohol and drive. Of those 492 fans, one lucky, randomly chosen fan won the "Responsibility Has Its Rewards" Sweepstakes, scoring a Miller Lite lounge set and two tickets to the September race in Richmond. NASCAR was sending a loud and clear message that weekend to thousands in the stands and millions who watched on television: Have fun, yes, but drinking and driving are not cool. What an ideal platform for such a message.

That's not the only "do-gooding" NASCAR has recently demonstrated. In the past few years, NASCAR started turning green, as in environmentally friendly green. In fact in 2011 NASCAR, Sunoco and the American Ethanol Industry launched a groundbreaking biofuel program that will clear a path for more widespread use of eco-friendly fuel in the future. NASCAR promoting sustainability? Yes, and not just in biofuel research and development.

NASCAR is planting trees, and lots of them, and in fact leads all sports in the number of trees planted through targeted efforts. The NASCAR Green Clean Air Program plants ten trees for each green flag that drops during races, an ingenious move that reportedly captures 100 percent of the carbon produced by on-track racing.

During the April 2013 month-long "Race to Green" initiative, more than 150,000 trees were planted by NASCAR, its partners, its drivers, and (a fact I find very cool) by its fans. Many of those trees were planted in areas impacted by natural disasters. The Virginia Department of Forestry has even partnered with NASCAR to reduce the sport's carbon footprint at tracks across the state. That's a responsible effort with the muscle of the Department of Forestry behind it.

Really putting its green where its mouth is, NASCAR is building LEED-certified (Leadership in Energy and Environmental Design) corporate office buildings. The twenty-story NASCAR Plaza in Charlotte and the impressive eight-story International Speedway Corporation and NASCAR headquarters building in Daytona are marvels of efficient use of solar energy, natural cooling, and recycled and sustainable materials. Who knew that this fringe bunch of beer-drinking, tobacco-spitting speed bandits has a social conscience, and that they're leading the way for other major sports as an example of being kind to the earth and the atmosphere?

It's not just NASCAR corporate that is spending time and resources on being green, either. Clint Bowyer, whom I now casually refer to as "my guy" (not sure if I have his OK to do that, but he is the driver for whom I'm pulling at every race), plants trees too. He and partner Martin Truex Jr. both with Michael Waltrip Racing, as well as David Ragan with Front Row Motorsports, showed up and pitched in at tree-planting events while encouraging fans to go out and do the same. Parks are being restored, and in one of my favorite green initiatives, United Parcel Service (UPS), partnering with NASCAR, donated thirteen thousand trees to be planted at the Flight 93 Memorial National Parks Service facility in Shanksville, Pennsylvania. I got teary-eyed when I read that, and I'm not kidding. I'm a sentimental mess when it comes to the sadness and tragedy of the events of September 11, 2001. Of course NASCAR is changing, and in very positive ways, but still the tradition of honoring heroes lives on.

I learned just recently that this year's FedEx 400, the Dover, Delaware, race that took place in June 2014, benefitted Autism Research, a cause that desperately needs funding to find answers to this puzzle that tremendously impacts families. The NASCAR Foundation presents the Betty Jane France Humanitarian Award to one NASCAR fan every year. The winner is a peer-nominated fan who demonstrates the ideals of charity and community that NASCAR Foundation Chairwoman France has championed throughout her lifetime. The Foundation gives a one-hundred-thousand-dollar donation to the children's charity of the winner's choice, along with a NASCAR experience of a lifetime. And the list of community involvement, help, and kindnesses goes on and on. I would think any fan would be proud of that type of change.

NASCAR partners have jumped right in with NASCAR's efforts to give back to the planet, children, and other deserving groups, including corporate giants such as Coors Light, Exide Batteries, Federal Express, Goodyear, UPS, Ford Motor Company, M&Ms, Sprint, and Toyota, just to name a few. NASCAR is flexing its corporate muscles, as well as those of its partners, to do good things. Yes, "yuppies" are all about being green, about being good to the planet, but is that such a bad thing? We are all leaving this planet to our children, whether they're NASCAR fans or not. Why not leave it a better place and have some fun doing it?

Another initiative I've studied, launched in 2000, is the NASCAR Diversity Internship Program. The program this year has selected and placed nineteen multicultural college students from universities across the country in ten-week paid internships that give the students hands-on experience in areas like marketing, broadcasting, engineering, communications, licensing, digital media, and other careers. The ultimate goal is that these young students will rise to take their places among the industry's next generation of key professionals. But again, this program is not characteristic of your mama's NASCAR. It's better, and there's room for everyone.

No matter how you slice it, NASCAR has changed, and continues to change, for the better. It is evolving, just as the fan base is evolving, and while this all means change, I am learning that change can be a good thing. As Geoffrey Norman so eloquently stated about NASCAR by way of quoting some Southerner he must have encountered along the way in his lifetime, "As they say in the South, 'Son, we ain't trash no more.'"

Chapter Fourteen
NASCAR's Domino Effect

Dollars seem to be drawn to NASCAR like moths to a Sunoco-fueled flame, if some of the numbers floating around out there are to be believed. From television rights contracts, to drivers' salaries, to car sponsorships, nothing, it seems, is done on a small scale. And what NASCAR does for the local economy when a race comes to town, well, let's just say it's a win-win for everyone. Just how much money are we talking about, where does it come from, and why is there so much of it?

In 2012, Fox Sports Media Group announced that they got the jump on their network rivals by extending their current contract with NASCAR. The existing contract, which still had two more years to go until it expired, has been extended for another eight years. That move seals the television rights deal between Fox Sports and NASCAR through 2022. In the new contract, Fox snagged the television rights to the Daytona 500 and next twelve Sprint Cup Series races of the season. The $2.4 billion dollar sweetheart of a deal breaks down to about $300 million per season. That reflects an increase of $80 million a year, compared to the contract previously in place. That's a lot of money, in anybody's book.

Sprint extended their sponsorship of the top race series through 2016, extending the original ten-year deal for another three years.

While the financial specifics were not disclosed by either Sprint or NASCAR (at least, I could not find the numbers in my research), the branding powerhouse obviously packs enough of a punch to convince both organizations to shake hands and do it again. I have no doubt that the handshake carried a hefty price tag.

ABC/ESPN jumped back into the business of broadcasting NASCAR races in 2007, and the sports network carries the heaviest load of races until the contract between them and NASCAR expires in 2014. In 2013, ESPN will broadcast the final seventeen Sprint Cup Series races and the entire NASCAR Nationwide Series. In what some called a surprise move (and what others called the handwriting on the wall), NASCAR inked a deal with NBC that will begin in 2015. NASCAR and NBC Sports Group announced in July 2013 that they reached an agreement that grants NBC Universal exclusive rights to the final twenty Sprint Cup Series races, final nineteen Nationwide Series events, select NASCAR Regional and Touring Series events, and other live content. Financial terms of the agreement, which will run through the 2024 season, were not disclosed by either party. One can only imagine the money involved in that deal. ESPN, the self-proclaimed center of the sports universe, carries the Chase for the Sprint Cup races and made it clear to all involved early on that they wanted to renew their deal with NASCAR. Turner, which currently carries six races in the middle of the season, was interested in adding more races to its schedule by most accounts. Some sources did imply that other networks, namely NBC and CBS, could possibly come knocking on NASCAR's door, jump-starting a bidding war that would definitely benefit NASCAR. Looks like they were right. Some analysts may think the television market value for NASCAR is slipping, based on the smattering of empty seats in the grandstands that's become obvious in the past few years. Whether the economy has impacted attendance, or the fans are disgruntled because of changes being made in the sport, television rights are still golden. Fox would not have been so eager to negotiate a new contract for more money, two years before the current contract would have expired, if their experts thought the market value was declining.

The contract amounts for billions of dollars are staggering, making it pretty clear to me that NASCAR's vital signs are looking very healthy.

The economic landscape surrounding the sport doesn't stop at network television, cable TV, and Internet rights negotiations. There are two other spinoffs of the financial strength and fan loyalty of the sport that I find very interesting—the local economic impact of a race coming to town, and the quirky, fascinating memorabilia market. When race weekend comes around in any of the towns that are home to a NASCAR racetrack, the local economy gets a healthy boost. Of course there are the obvious benefits of thousands of fans flocking to town, renting hotel rooms, eating in restaurants, and shopping. I also learned something I didn't know before, but it makes perfect sense. A NASCAR race infuses local economies with much-appreciated cash. Locals are hired when a race comes to town, taking jobs that range from security, to ticket-takers, to other event staff. Local groups are often given the opportunity to work either for vendors or the stadiums, earning money for their charities, school groups, or churches. Consider, for example, the annual NASCAR race event at Las Vegas Motor Speedway. In that one spring weekend, about $240 million are pumped into the Las Vegas economy. Hotels, shows, casinos, and restaurants are beehives of boisterous activity, as an estimated 70 percent of race weekend fans are out-of-towners. Locals scramble to snag any jobs associated with the race: extra bartenders, skycaps, valets . . . the list goes on and on. Las Vegas Motor Speedway pays for additional police officers, firefighters, and traffic control measures. NASCAR race weekend has become one of the biggest annual events in the city, a pleasant surprise to organizers who, back at the event's 1998 inception, feared that fans would show up, go to the race, then leave without spending much money in town.

Here's an interesting fact: during race weekend in Las Vegas, the speedway property balloons to become the fourth-largest city in Nevada. Local charities pocket somewhere in the neighborhood of $130,000 for their efforts, and hundreds of airmen from nearby Nellis and Creech Air Force bases volunteer to work, with their units being paid for the manpower. The bases then use that extra money for activities for the airmen. When race day comes to Vegas, everyone wins. NASCAR track cities across the country have similar success stories to tell; from Dover to Atlanta to Sonoma, billions of dollars have been infused into local economies over the years because the boys and girls of NASCAR come to town.

Now, on to a topic that absolutely fascinates me, and it has for years: The NASCAR memorabilia market. Many years ago, I worked not only as a freelance writer, but a freelance website designer, as well. Mind you, this was well before the days of software packages and programs that made designing websites fun and fluffy. I got into the business quite by accident, and it was back when the designer actually wrote the HTML code. There were no colorful buttons to click marked "align" or "publish." Back then, website design was a real nerd's playground. Anyway, I was hired by a company that sold NASCAR memorabilia out of a real brick-and-mortar store in, of all places, North Carolina. This company was one of the businesses early on that decided to venture out into the scary, untested world of the Internet, previously occupied solely by the US government. When the store owner contacted me and we began discussing his business, I remember very clearly thinking, *NASCAR memorabilia? Are you kidding me? Who in the world is going to buy that, and on the Internet, no less?* But, the man was willing to pay my asking price, and I got down to business building his website. That was my very first introduction to NASCAR collectibles and memorabilia and to the people who buy the stuff. Incidentally, this man's website was by far the most popular, heavily-trafficked site I ever built. In the first year alone, Internet sales topped $1.6 million, and this was back in the early 1990s. I will never forget some of the item photos that crossed my desk way back then, and today the selection of merchandise is just as intriguing.

NASCAR fans are a different breed. They are devoted, diehard, loyal, ravenous people who love the sport, their driver, and anything that is associated with both. There is a Midwestern Mom blog about NASCAR and unusual memorabilia this woman uncovers at flea markets (and crushes she apparently has on particular drivers).

There is a blog that invites newlyweds with a shared love of NASCAR to tell the world their love stories. Fans love NASCAR. Of course they buy T-shirts, hats, and trading cards; what sports fan doesn't? But those tried-and-true items are just the tip of the iceberg. Collecting and selling die-cast stock car replicas, either 1:18 or 1:24 scale models, is an industry all by itself. According to an August 2010 article in the *New York Times Business Day* written by Dave Caldwell, sales of these tiny, intricate replicas was estimated to be about $250 million annually in 2002. As NASCAR has been affected by a sputtering economy, so have the sales of these miniature cars. With sales down by about a third just five years later, the bottom line is that industry was at the time very obviously tied to NASCAR's overall experience—the first signs of dwindling attendance in the grandstands. Still, NASCAR logged an impressive $1 billion in merchandise sales in 2009, and one quarter of those sales included officially licensed die-cast car replicas. The tiny cars are incredibly detailed down to the tiniest features, and they are not toys. Selling for around $60 each for current models, they are really meant for display only. Many are in circulation now that have been in families for decades, being sold to a market of eager buyers,perhaps by grandchildren who have a greater need for money than a passion for NASCAR. Interestingly, but not surprisingly, sales of a particular driver's replica car rise in the days following a win. What I find interesting is the outlandish ideas that NASCAR memorabilia dealers come up with and better still, the fact that people buy it. Big-box retailers, such as Target, Wal-Mart, Sears, and JC Penney sell NASCAR memorabilia, and so do little mom–and-pop businesses run out of basements throughout the country. There's a market for it out there, and to me, it seems the weirder the better as far as marketability is concerned. Let me give you some examples of what I mean by weird and outlandish merchandise. How about outfitting an entire bathroom in your home with NASCAR everything—shower curtain, toilet seat and cover, towels, rug, and toothbrushes? Not enough NASCAR for you? You can custom order (or download the plans to make it yourself) a NASCAR bathroom door. If you're looking for a little something special for your Christmas tree this year, how about a NASCAR-themed sleigh with Jeff Gordon's #24 painted on the side? I could go online right now and buy myself a Clint Bowyer handbag, if I took the time to wade through all the other merchandise to find it. I can buy a mailbox cover with any driver's number I choose. I can buy collars, leashes, T-shirts, food bowls, and toys for my dogs, all NASCAR or driver-themed. I can buy a piece of a racetrack. I can hire a wedding planner to orchestrate a NASCAR-themed wedding, though heaven forbid I should ever need that service again. I can even buy a Dale Jr. pinwheel yard spinner lawn or Joey Logano lawn flag. I looked for something really crazy, like someone selling a piece of toast with the face of Brad Keslowski looking back at you if it's held at just the right angle in just the right light. Didn't someone sell a grilled cheese sandwich on eBay that supposedly had Jesus' likeness on it? I never found anything like that, at least not NASCAR-related. I located a NASCAR memorabilia store—a real one, not an online store—not far from my house, and I went to check it out for myself. Oddly, I must have passed this store a hundred times before, as it's discreetly nestled between one of my favorite lunch restaurants and a popular neighborhood supermarket. The very narrow storefront does not call attention to itself; you'd really have to be looking for the place even to notice it's there. Frank, a friendly guy who's probably in his early seventies, owns the store. When I met him, he was wearing Dale Earnhardt Sr. suspenders. I wondered if that special touch was for my benefit, as I had called him a few days before we met to ask for a bit of his time. I've come to the conclusion that the suspenders were not at all for my benefit; rather, I suspect that they're just part of his daily wardrobe. The walls of his shop were covered with Dale Sr. posters, signed memorabilia, and apparel, none of it for sale. These were part of Frank's personal collection. Shelves along the long aqua-blue walls displayed jiggly bobblehead dolls, tightly packed shoulder-to-shoulder and grinning happily at whomever happened to be looking them over. A large plaque clock at the far end of the store proudly displayed the name of Jeff Gordon. Glass cases housed individual trading cards, some signed, and some more than thirty years old. The most expensive card in the case was priced at $1,200 and had been signed by both Bobby and Davey Allison. Inside the cases with the cards were drivers' gloves, a signed helmet that had belonged to Richard Petty, and various other items that needed a bit more protection than a shelf offers. Pieces of actual race cars—a hood here, a bumper there—were propped along the back walls. Frank was sipping on a Dr. Pepper when I walked into his shop, and he offered me one as we sat and talked about how he got into the business of peddling NASCAR memorabilia. "I've been a racing fan since I was eight years old, when my daddy took me to my first race down in Hampton," Frank said, referring to Atlanta Motor Speedway when he spoke of "Hampton." Back then, before the spaghetti-bowl network of highways that connects north and south in Atlanta was built, traveling from Dahlonega, Georgia, where Frank grew up down to Hampton was easily a daylong endeavor. "I had listened to races on the radio with Daddy, sitting at the kitchen table on Sunday afternoons, but I had never seen a race up close. I never forgot it, either. Loved it ever since." In fact, Frank met his wife of nearly forty years, Eileen, while they were both at a race in Charlotte. "She was there with her family. We kept looking at each other, and I kept motioning for her to walk down with me to get a bottle of pop. She finally did," Frank shared, reminiscing with me as he sipped from the sweating can of soda. The heat and humidity that day were soaring, and Frank apparently doesn't believe in air conditioning. He and Eileen still go to as many races in both Atlanta and Charlotte as they can. They raised their two sons to have a love of NASCAR, and now, they take their grandchildren to races with them. In 1990, the couple opened their first memorabilia store in Georgia. When I asked him what the strangest piece of memorabilia was that he had ever sold or traded, he said that he really couldn't remember having sold anything that was out of the ordinary. He did, however, remember one woman's request of him, and she made it not too long after his first shop opened for business. "This woman came in one day, first thing, when we opened the doors for business, and she asked me to keep an eye out for two things for her. She wanted either hair or whiskers from Junior

Johnson, and she wanted any gum that any NASCAR driver had ever chewed. She had heard that Junior's whiskers were out there somewhere. I don't know why she wanted chewed gum, and I didn't ask her." I couldn't imagine why she'd want them, either. Frank showed me a few other pieces of his prized offerings before we wrapped up our conversation. I held a piece of the track from Daytona International Speedway in my hands, and it was priced at $125.00. There was a used connecting rod in its own display case ($75.00) and a framed photograph of Lee Petty (Richard's dad—$160.00). Before we parted ways, I asked Frank whether the plethora of Internet memorabilia sites, which must number in the hundreds by now, has hurt his business. While he answered that yes, a lot of people opt to shop at home, scrolling and clicking to add to their precious collections of cool NASCAR stuff, there are still those tried-and-true customers who like to see and touch items before they buy them. They like to check in periodically to see what he has to offer for sale that's new. They also like to see what's sold since their last visit. Most of them like to stick around a while and talk racing, whether it's about a race that took place thirty years ago or the one that was run just last Sunday.

There is no accurate way of telling just how much money changes hands annually with respect to the sale of all NASCAR memorabilia, as there is obviously no central reporting or tracking system. It's a multi-million-dollar industry, no doubt. And it has me wondering, just how much does a wad of chewed gum go for these days?

In one of the strangest NASCAR memorabilia stories I ran across during my research, former NASCAR driver Jack Ingram's grandson was arrested in 2011 for stealing about $5,000-worth of his grandfather's keepsakes. The young man took the diamond ring his grandfather won for his 1982 Busch Series championship, another 1985 championship ring, a watch, and his grandmother's high school ring. After stealing his grandfather's belongings, the grandson reportedly called Ingram anonymously to demand $1,000 for their return. I found it funny that the kid undervalued the items like he did and that Ingram refused to pay, anyway.

Much of the memorabilia I ran across during my travels this year had ties not just to the racetrack, but to the pits along the track, as well. Everything from tools to lug nuts to pit crew gear is offered up for sale on various websites and in quirky little memorabilia shops scattered across the country. Why the fascination with the crews and their jobs that, in my early, shallow understanding, were merely support roles for the glamorous drivers? Weren't they merely considered behind-the-scenes workers, doing jobs that anybody could do with very little effort? As I stated earlier, my understanding way back then was shallow and way off the mark.

Chapter Fifteen

"I Need to Make a Pit Stop"

So often, the face of any sports team is really the faces of the key players. In the NFL, teams are characterized by the quarterback, his style and personality. In Major League Baseball, the pitchers are often credited or blamed for the success or failure of the entire team. NASCAR is no different. To many spectators, the face of NASCAR is, in fact, the drivers. While the driver may be the most visible member of the team, he is by no means the only member. There is an entire cast and crew on every team, and each person has a job. The team's pit crew demonstrates that premise clearly, and it's at a race that their speed, strength, and magic are most evident. I suppose I knew that every team has a crew that takes care of car maintenance at a race, but that's pretty much all the thought I ever gave the matter before I decided to really learn about NASCAR. As it turns out, the work that the pit crews do is intricate, complicated, and physically demanding. The six- to seven-person crews are actually a team of athletes competing against other teams, with the fastest team being the winner. In fact, up until 2013, there has been an annual competition called the Pit Crew Challenge, a staple for the past eight years at the Sprint Cup All Star Weekend in Charlotte. The 2013 event was cancelled for lack of a sponsor. Sprint was the most recent sponsor, but the company did not renew the contract after 2012.

Once racing teams saw that a top-notch pit crew could actually give them the winning edge on race day, more attention and funding were devoted to enhancing performance in the pit. Teams began wearing special uniforms and equipment, even down to their gloves. Gone are the days of crew members wearing knee pads designed for volleyball players, baseball batting gloves, and off-the-rack athletic shoes. Mechanix, the company that makes different types of gloves for the jobs on pit road, works with crew chiefs, jack men, tire changers, and carriers to design gloves that enhance performance and speed. The gas man's gloves are different from the tire changers' gloves, which have extra reinforcement at the knuckles to protect them from the inevitable scrapes against the tire rim. The tire carriers' gloves have a grip similar to an NFL wide receiver's gloves to prevent slipping. Every second saved in the pit is essentially another second handed to the driver.

The precisely choreographed drills that are executed in the pit require a lot of practice, focus, and determination because again, everything about NASCAR boils down to one thing: speed. I might also add that, in my opinion, the pit crews are every bit as fascinating to watch during a race as are the cars. When the team executes its precise drills as the driver enters the pit, the performance is no less than perfect, at least from the fans' perspective in the stands. I learned, however, that sometimes things can go a little wrong or very wrong, and that's when the crew's finesse is really tested.

David Ragan's pit crew chief Jay Guy explained to me how a crew is selected and what's expected of each member during the week leading up to a race and during the race itself. "There are six guys who go over the wall at a race," Guy said. Those six guys (or women) can be part of the racing team that works on the cars at the shop, or the pit crew members can be selected separately by the pit crew coach, who's constantly got an eye out for new talent. "You have to be an athlete, first and foremost," said Guy. I asked if it mattered to him whether a crew member was male or female. "Is mixing genders in the pit a distraction?" I specifically asked, then used as an example Christmas Abbott, who shadowed Michael Waltrip's racing crew at the Daytona 500 this year and has actually signed a deal to be a pit crew member with NASCAR's Camping World Series. Under the newly inked contract, Abbott will be changing tires for female driver Jennifer Jo Cobb. "You have to have great hand-eye coordination. You have to be athletic. You have to be able to think fast, especially when a problem arises in the pit. If you can do all that, I don't care who or what you are," Guy told me. It's not unusual for proven athletes in other sports to be hired as crew members, once they add the necessary mechanical training to their athletic skills.

"Everyone has a specific job. Our crew practices every Tuesday and Wednesday of every week," Guy explained, going on to tell me how the team is constructed. "The jack man is my quarterback. He keeps an eye on everything that's going on, while the tire changers are focused on doing their thing at either end of the car." That thinking makes a lot of sense, as the jack man lifts the car off the ground in a few quick movements, standing exactly in the middle of the car as he does so. From that perspective, he can see what the team members on both ends of the car are doing at all times. Keeping in mind that a standard pit stop takes about twelve to thirteen seconds, there's a lot to watch and not much time in which to watch it. Guy estimated his current crew, for a four-tire change and fuel, averages between 12.4 and 12.6 seconds per stop. That's lightning fast.

Here's how the jobs in the pit are divided. If you've ever watched a crew do its thing during a race, I guarantee that it will take longer to read the job descriptions than it will for the crew to perform them, even if you're a fast reader.

The jack man jacks up the car so that the tires can be replaced. He (or she) usually pulls the old right rear tire off of the car after the rear tire changer loosens the lug nuts. This helps get the new right rear tire on faster. This guy is the one who signals for the driver to leave the pits by dropping the jack. It's really something to see, with the tires already spinning as the car drops back onto the road, literally burning rubber as they make contact with the surface. It's up to each team member to be sure he's out of the way when the car drops. If something should happen—someone slip up or be off just a second or two on their timing, the results could be disastrous.

Pit crew in action at Darlington.

The front and rear tire changers change only the front or only the rear tires, and they use an air wrench to do it. The tank that powers the tool, brought in along with hundreds of other tricks of the trade in the team's pit box, spews a fine mist into the air as the lug nuts are spun off, then back on in a matter of a few seconds. At a night race, the mist rolls out over the track like the fog in a Sherlock Holmes tale, only it's gone almost as quickly as it appears. This is NASCAR, after all. Fast.

The gas man fills the car with gasoline by using a special gas can. I can best describe the appearance of the can as a giant red water bottle, the kind that attaches to the side of a gerbil cage. He lifts it up high, fitting the nozzle into the gas tank. Don't blink, or you'll miss the entire process. The gas man may also help pull tires off the car if he finishes his job before everyone else does.

The front tire carrier carries the new front replacement tires over the pit wall, then guides them onto the studs. He rolls the old front tires to the pit wall after the front tire changer pulls them off of the car. The front tire carrier is usually responsible for clearing debris off of the grille of a race car and/or pulling the front fenders away from the tire if necessary. Drivers are constantly bumping and nudging each other, or sometimes they're even bumping the wall. This member may also be responsible for adding tape to the grille during a pit stop. That last responsibility has something to do with front end downforce, which sounds awfully important. I believe it has something to do with ergonomics.

The rear tire carrier's job is bringing new rear tires over the pit wall. Typically, it works this way: On the right side of the car (farthest from the pit wall), he's responsible for guiding the new tire onto the studs, making any necessary adjustments to the rear track bar and/or wedge (remember the wedge adjustment?), and carrying the old tire back to the pit wall. On the driver's side of the car (closest to the pit wall), he is usually only responsible for sliding the new tire onto the studs. A seventh person is sometimes allowed over the wall, to clean the windshield, and sometimes to attach dark shields to improve the driver's vision as the sun begins to set on the track. Races at Atlanta Motor Speedway, for instance, were known for presenting glare issues when the sun set on a race that would run into the night.

There is a tremendous amount of responsibility on the pit crew's collective shoulders, especially those of the crew chief. If anything should go wrong—if the engine or body deviates at all from NASCAR specifications,—not only can points be deducted from the driver and owner, but also hefty fines can be levied against the driver, owner, and crew chief. The 2013 season saw a few examples of that very thing.

The crew chief must coordinate and oversee every detail of the car and the pit process with hawk-eyed precision. Helping him do that is the traveling garage called a pit box.

Every team has one, and every crew member has a specific job once the box arrives at the track in the driver's pit. That job may be to put certain tools in certain drawers or cabinets. It may be to hook up all the electronics. It may be to place the driver's numbered flag on the long, thin, flexible pole that overhangs the pit, its sole purpose to help guide the driver to the proper pit without taking too much precious time to figure out which one is his. Whatever the jobs are, they are done quickly and quietly. The box itself looks

Driver Carl Edwards pit box, the "toolbox" taken to every race.

like a giant toolbox with seats on the very top. It has every tool
(some mechanical and some hand tools) that could possibly be
required during a race, and on top, the box looks like the CNN
newsroom. There are flat-screen TVs, a satellite dish, computers, and
seating, with canopies to keep everyone dry and comfortable in
case of rain. These computers and other electronics help the chief
and his crew monitor everything that goes on during the race, from
the race itself to the data communicated by both the car and the
driver. This information helps the team foresee potential problems
and head them off before they become actual problems. The steady
stream of data ultimately helps the team shave, literally, milliseconds
off the time it takes to pit the car. Those milliseconds can and often
do mean the difference between taking a trip to Victory Lane with
your crew and a few bottles of champagne, and taking second, third,
or even tenth place in a race. I think that if push came to shove,
the crew chief could actually perform surgery on the driver if he
had to, using the pit box and all the high-tech tools safely tucked
away inside, ready for action at a moment's notice. The pit box
also provides more precious real estate on which to display those
ever-important sponsor logos.

Dubbed "pit stops" because of the deep pits in which mechanics sometimes stand to work on cars, the pit stops of yesterday hardly compare to the twelve-second magic shows of today's NASCAR. Really, more than six decades ago, pit stops were considered to be afterthoughts, necessary evils. Having to tighten the leather belts that held the doors shut, or having to refresh the white shoe polish that marked the numbers on the top and sides of the cars for a hundred miles on a dirt track, were a far cry from keeping a gas-hungry engine and beat up body race-ready for five hundred miles.

Innovation legend and recent NASCAR Hall of Fame inductee Leonard Wood pioneered many of the techniques used today that helped revolutionize pit road. The chief mechanic for the famous Fords routinely featured by Wood Brothers Racing, a team competing in NASCAR as far back as 1953, Wood was on the scene when crew chiefs made the switch from tire irons to power guns to remove the lug nuts from tires in the pit. That shaved about twenty seconds off the standard forty-five. Next, the Wood brothers zeroed in on the jack and how to speed up the process of lifting and lowering a car. Three pumps now, and the car is high enough off the ground to exchange its tires. Then Wood and his crew streamlined the fueling process, utilizing gravity to pour the gas into the tank instead of a hose to pump it. Then they designed quick-pull handles for the gas cap to speed things along, easily shaving two seconds off their time. Before long, Leonard and his brother Glen became known throughout the sport as the fastest pit crew around. In 1965, Ford Motor Company asked them to join driver Jimmy Clark's pit crew for the 1965 Indianapolis 500. Even though the brothers found themselves working with an unfamiliar all-British crew and had little time to gel as a team, their know-how and common sense paved the way to Clark's victory. Many say that the importance of the pit stop shifted from the back seat to the front at that race in 1965.

Increased speed brings with it the increased probability of an accident, not only in the driver's seat but also on pit road. The job of the pit crew does not come with any guarantees of safety; in 2009, NASCAR fans and officials watched in frightened amazement as a mechanic from Marcos Ambrose's crew chased a wayward tire across the racing surface and out into the infield grass. Probably thinking he was doing a good deed and trying to prevent a crash caused by the loose tire, the mechanic's actions brought the race to a halt and likely affected the ultimate outcome of the race. Everyone was shaken up, including the drivers.

Tragedy struck Bill Elliott's championship-winning pit crew in 1990, when tire changer Mike Rich was killed in a freak accident while on pit road. Elliott's car was already pitted when driver Ricky Rudd entered the pit, spun out, and pinned Rich between his car and Elliott's as fans and team members looked on in horror. That fatal accident is directly responsible for speed limits being imposed on drivers entering pit road.

The lower speed limits have undoubtedly led to fewer tragedies, but in 2007 another serious accident befell a crew member. Michael Waltrip's gas man, Art Harris, was hit in the head by a bouncing tire during a pit stop. While Harris eventually made a full recovery, he spent time in the hospital as a result of the head injury. That mishap led to NASCAR's still-standing rule that tires cannot be rolled to the wall after removal. They must be carried.

While the average annual compensation for pit crew members is as closely guarded a secret as NASCAR salaries and purses (I'm not sure anyone knows the exact numbers for crew members or drivers), I have heard estimates that range from about $70,000 to more than $100,000 a year, before bonuses are granted for speed. That's not bad money. Of course, there is an awful lot of travel involved, and there is the possibility of a dangerous accident to consider that comes along with the job.

I was curious enough—fascinated, really—with the whole process that takes place from A to Z during a pit stop, that I had to figure out how to try it myself. As ridiculous as that may sound (after two knee surgeries, I am best timed with a calendar when it comes to doing most anything physical), I figured out a way to do it. On my first visit to the NASCAR Hall of Fame near the beginning of the 2013 season, I saw that there was an interactive area of the displays that gives visitors a chance to test their skills in the pit. Once I really dug into the research and had a few races under my belt, I decided to go back to North Carolina and try it myself. I had to know what it was like, even if it would be under the friendliest, safest conditions.

My husband and I took off for Charlotte for the second time about mid-May. While the heat and humidity have usually settled over the South like a wet wool blanket by that time, cooler weather kept making encore appearances that year. We lucked out and had a bright, sunny, cool weekend on which to travel. As always, the sights and scenery on the drive did not disappoint. We waved to the Gaffney peach as we zipped past it on I-85, my husband telling me that I had forever ruined the sight of it for him by comparing the peach water tower to a giant human derriere. For whatever reason, we saw not one, not two, but three instances of people having pulled off to the side of the road to, well, use the facilities where none existed. Yes, whether the culprit was a man or his young son, it seems that it is becoming more socially acceptable to do that in plain view of travelers. I could never relax enough to do that, but I'm not a guy. I snapped a picture of the last guy we saw doing it, and if I'm not mistaken, he waved to me as I took the photo.

We stopped at a diner just outside of Greenville, South Carolina, for lunch. I just love diners, and I frequent them as often as possible as much for the people as I do for the food. There's something about the atmosphere in a diner that encourages conversation. I would never strike up a conversation with a complete stranger in one of those loud, busy restaurants that advertises on television and duplicates floor plans and menus across the United States. But a diner, with its simple menu, usually a counter lined with shiny chrome seats and topped with red pleather, and with day-old cakes and pies proudly displayed under glass domes, is just a great place to hang out and talk. I think I developed a love for diners and good conversation by being a newspaper reporter. I have conducted many an interview over a cup of coffee and a slice of lemon meringue pie.

Our waitress at this newly remodeled diner was a particularly friendly woman named Gloria, who in a few short minutes shared with us that she is twice married and twice divorced, with two teenagers living at home. Her first husband, who by her own account is a "dog" who never paid child support or lifted a finger to raise the children, was moving to New Mexico in about a week, a clever attempt to escape his second wife, who also happens to be Gloria's best friend. Her second husband, an all-around good guy who just couldn't hold a job, still spends as much time on her sofa as he did before the divorce. "I just can't get rid of him, bless his heart," she confided, adjusting her pencil behind her ear and whispering to us as though we were lifelong friends.

After filling us in on all of her recent business, Gloria brought our two glasses of iced tea and efficiently whipped out her order pad to take our lunch order. She volunteered that the day's special, back by popular demand, was fried bologna sandwiches. My husband's eyes lit up like a Christmas tree. You see, fried bologna sandwiches are as beloved a Southern delicacy as crab cakes are in Maryland. I don't believe I've ever heard anyone else besides a Southerner ever admit to eating one, come to think of it. And I know for a fact that bologna is rarely featured on a menu, even in the South. I think it's a health code violation. As I said, although my husband hails from Ohio, he integrated seamlessly with the Southern food culture some forty years ago. He gets tears in his eyes at the mention of fried bologna sandwiches, moreso since I banned the "Southern round steak" from our home as soon as we married fifteen years ago. It's not that I'm too proud to eat it; I ate enough bologna sandwiches as a kid to feed a small country, I'm sure. It's just that I have no idea what's in bologna. As an adult, that matters to me. As a mother, that really matters to me. Even the stuff packaged with labels that read, ALL NATURAL or ALL BEEF makes me nervous. Chlorine is natural. Hooves, as long as they were attached to a cow at some point, are still technically considered to be beef. It's just not natural to eat a substance that takes on the shape of whatever packaging it's in, and for which the ingredients list reads like a science experiment. Something else that bothers me about bologna is that when it's fried, it shrinks and curls up. The fat content has to be at least 50 percent, because the piece of bologna shrinks by 50 percent when it's fried.

Oh well, the curled meat sandwiches were on the menu, and there was no getting around it. My husband ordered two of them without waiting to see what his other choices were or what I cared to have for lunch. I ordered a chicken sandwich and was promptly told "Naw" by our waitress. "Pardon me?" I asked, thinking that I did not hear her clearly. I thought she just told me I could not order a chicken sandwich. "Naw," she repeated, and while she scribbled something on her order pad, she informed me that I, too, would be having a fried bologna sandwich for lunch. "But I don't like bologna," I replied, instantly feeling like a child trying to wheedle her way out of eating Brussels sprouts. The man sitting in the booth behind me turned around and interjected his thoughts on the matter. "You ain't never had a fried bologna sandwich like Fred makes 'em. Get the bologna." *Did a complete stranger just tell me what I'm having for lunch? Yes, he did.*

I opened my mouth to object from another angle this time, but Gloria was telling me that I simply could not miss this opportunity, as who knew when the sandwiches would be pulled from the menu again. Before I could ask why they would be removed from the menu, she was off and running, shouting out for the entire diner to hear that we would be lunching on bologna sandwiches, with chips and a pickle, for lunch.

The man sitting behind me assured me that I wouldn't be sorry, and his wife backed him up on that. My husband assured me that, if I didn't like my sandwich when it arrived, he would be happy to eat it for me. When our plates arrived just a few short minutes later, Fred himself waved his dripping spatula at me and winked, yet one more affirmation that Gloria had made the right choice for me.

My diner tale, to make it short and sweet, ended with me eating my bologna sandwich. I took my first bite strictly out of peer pressure and hunger. I finished it, because that first bite opened up a floodgate of childhood memories that made it taste simply scrumptious. I can say it now; I enjoyed it. While I will still never buy bologna and bring it home to stash in our family's refrigerator (the expiration date is usually about a year or so out from the date of packaging), the sandwich I ate that day in that little diner in

Greenville was delicious, in more ways than one. My husband finished both of his in no time flat, and I caught him a time or two running his finger along the plate and surreptitiously licking it, sopping up the grease that had escaped the bread and dripped onto the plate, already congealing there. Bologna does that to you too. It makes you lose all sense of manners and propriety.

When we paid our tab and left, everyone—and I do mean everyone—in the diner yelled their goodbyes and waved, even Fred back in the kitchen. Gloria, of course, hugged us. I do love a small town diner.

We arrived in Charlotte in the late afternoon. I suppose we had enough time to make it to the Hall of Fame that same day, but I wanted to have a restful, calm evening. I wanted to get plenty of sleep and have a good breakfast in the morning, because I had one purpose and one purpose only for going back to the institution: to try my hand at changing a tire and fueling up a car, racing against the clock. I wanted to give it my best shot.

On Saturday, we arrived at the Hall of Fame the minute it opened for business. I wanted to be sure to beat the crowd that would inevitably make its way to the interactive pit stop exhibit on the top floor. Even arriving that early, I had to patiently wait my turn as two young boys of about twelve or so were already waiting for the lady at the exhibit to get her timer set up. She actually times each participant, something I remembered from our first trip. As I waited for the two boys to finish up, I rolled my neck and rotated my shoulders, shook out my arms, and flexed my knees (as far as they flex these days, anyway). I was behaving as though I was getting ready to take the field at the Georgia Dome in Atlanta, leading the Falcons to a victory. I must have looked ridiculous, but I didn't care. I was on a mission.

A soon as the two boys finished testing their mettle in the pit, it was my turn. I took a couple of deep breaths, shrugged off my husband's admonishment to be careful and not hurt myself, and stepped up to the lady with the timer. I ignored her polite but concerned question: "Are you here with your grandchildren?" "No," I answered, curtly informing her that I was here to change a tire and fuel up the car myself. I also ignored her slightly raised eyebrow prompted by my response. Opting to fuel the car first, I rolled my shoulders one more time, looked at the woman with the timer, and said, "Go." The tank is heavier than it looks, and getting the nozzle fitted perfectly into the gas tank is harder than it looks. While I felt as though I was moving like the wind, I put up a solid two and a half minutes on that task. The next job was to change a tire on the car. For the record, the tires are heavier than they look too. Using the air wrench, I zipped through the lug nuts and pulled the tire off, dropping it on my foot and uttering an unintended expletive, which drew another raised eyebrow from the timer lady. After wrestling with the tire and tripping over the air hose, I finally got the holes lined up, heaved the tire into place, and tightened the lug nuts to secure the whole mess. Since I'm being perfectly honest about my performance in this undertaking, I logged a cool four minutes flat on that job. There was a line forming at the exhibit, with antsy kids looking at me in utter disbelief, shifting from one foot to another with obvious impatience. My husband was looking at me with a combination of pity and embarrassment that I've only seen on his face a time or two during our entire marriage. He continued looking at me that way as he explained that the tire used in the exhibit was more than likely a lot lighter than an actual tire on an actual race car, and he knew for a fact that the gas tank was lighter. "Are you sure?" I asked him. He patted my hand as he smiled a barely concealed smile. "I'm pretty sure, honey." The day's activities were pretty well summed up when he said brightly, "Look at it this way. At least you didn't hurt yourself."

We spent the rest of the day perusing exhibits that I had either missed during our first trip or, more likely, that I didn't fully appreciate the first time around. I paid particular attention to the items that told the tale of the NASCAR of yesterday, glimpsing a little piece of the back story here, another there. On display is the original 1950 seating chart for Darlington Raceway, designed by founder Harold Brasington himself. There is a 1952 army surplus radio on display, driver Al Stephens's first attempt to communicate with his team during a race. We saw Jeff Gordon's original application into NASCAR. And there's Glory Road, the zero-to-thirty-three-degree-banked turn, with eighteen cars displayed on it that hold some pretty significant spots in NASCAR history.

When we left the Hall of Fame late that Saturday afternoon, I have to admit, I was a little surprised and a little bummed out by my turtle's pace in the simulated pit. I chalked it up to the pressure of having all those people watching me, and the lady with the timer, who had to be off her game as she clocked my painstaking efforts. Twelve seconds in a real NASCAR pit? Really? The more I thought about it, the more I craved the comfort food of my childhood: a fried bologna sandwich. And that brings us around to the topic of food, partying, and tailgating, as only NASCAR fans can do it.

Chapter Sixteen
Tailgating, NASCAR Style

Of all the things I've experienced in my NASCAR-fueled travels this year, I have to say that tailgating has by far been the most fun and the most entertaining part of the research. Anyone who's ever tailgated before a ballgame knows that unmistakable, anticipatory party vibe well. It's a jovial one in which everyone gets along, and laughter, conversation, music, and perhaps a ballgame on the radio are the background sounds that season the whole mix. Complete strangers not only speak to one another, but rather they pull up a chair and really talk with one another, always finding out sooner rather than later that they have something in common. I am naturally drawn to any environment in which people are comfortable enough to lower the walls of guarded pretension for a while and really talk to one another. In a relaxed and festive atmosphere such as tailgating, we get to see people for who they really are. Genuine conversation and unaffected laughter are probably my favorite sounds in the entire world.

And the smells—oh my, the smells. There's that familiar scent of family-recipe hamburgers sizzling alongside juicy hot dogs on a grill. There's bratwurst marinated in a secret sauce, the recipe passed down from one generation to the next like a well-guarded secret. There's barbecue chicken, the sticky-sweet, gooey sauce cooking to a crisp black in some spots, a sure signal to the cook and ravenous onlookers that it's ready to eat. There are shrimp, painstakingly strung on skewers and reeking of garlic and butter, dripping and spitting into the fire with tantalizing slowness. There are thick, juicy steaks that sizzle and smell simply divine. If it can be cooked to perfection over an open fire, you'll find it sizzling and calling your name at a tailgating party.

Even the grills and smokers are often unorthodox contraptions at NASCAR tailgating and campsite throw-downs (that's Southern speak for "parties"). Some that I saw had been hauled to the site in the bed of a pickup truck or strapped onto the back of a truck or a camper. But still others had been permanently welded onto the back of pickup trucks or trailers. I saw many of those metal vehicle/ grill creations in my travels, as well as big black smokers with tall, round chimneys, those chimneys charred even blacker by exhaling years of slow, wood-flavored smoke. You know a person's serious about cooking over fire when they've made a grill or smoker a permanent part of their transportation.

One Bristol tailgater with a flair for the unusual actually had a smoker in the shape of a well-endowed woman custom-made (think those "naked lady silhouette" mud flaps you often see on eighteen-wheelers). He told me he had paid $16,000 for the welded artwork, a salute to the female physique expressed in curvy black metal. I'll leave it to your imagination as to where the chimneys were located. It was interesting, yes, but seeing this man cooking on what he called "the lady," alongside his preteen son and his wife, seemed a little weird to me. "The lady" was fashioned to be improbably out-of-proportion, but I will say this: the guy could work some magic with a pork butt roast. His sweet, slightly charred barbecue was nothing short of exquisite, brownish-black on the outside and a tender brownish-pink in the center. The twelve-pound chunk of meat had been smoking on a bed of coals and hickory chips (and some other flavor of chips, the source of which he refused to disclose) for some thirty-six hours when we met him at that race in late August. He and his little family talked openly and excitedly with me about my NASCAR debut adventure, and in no time at all, he was gingerly pulling off delectable chunks of sizzling pork heaven and sharing with my friends and me. I have no compunction whatsoever about taking meat from strangers at a tailgating party; we ate the shredded pork and washed it down with an Alabama Slammer, a yummy and seductive mix of vodka, Southern Comfort, amaretto, sloe gin, and orange juice. It was, after all, only 10 a.m., so the orange juice somehow validated the Slammer as a breakfast drink. Yes, I like this NASCAR tailgating thing very much.

Every single race I attended in my season of travel had a tailgating personality all its own. Some were wild parties (Talladega), some were more family-oriented affairs (Darlington), and some were amazingly upscale occasions (Daytona), more like outdoor cooking shows than prerace parties. But every one of them was an education, an experience, and a new gastronomic escapade. Many NASCAR tailgaters take the process to an entirely new level—one that marries art, science, culinary skill, and sometimes even sex, with high school shop class.

I have one memory of Talladega tailgating that stands out: I walked with my husband and son past a group of guys who looked as though they had been camping in their tents for at least a week. Their clothes were wet and dirty. Their tents were waterlogged, and their big pickup truck was mired hubcap-deep in that clay-like, gloppy mud that only the southeastern states can cough up after a long period of rain. Even though those guys looked as though they should have been miserable (I'd be packing to go home to civilization), they seemed to be having a great time. Not the type to let a little rain ruin their party, they were tailgating. They had a game of cornhole set up, and the four players were good-naturedly trash-talking one another about who was going to win. They were cooking too, and they were doing it on one of the most ingenious grills I do believe I've ever seen. The men who were in charge of cooking what appeared to be hamburgers were gathered around a pit that had been dug into the ground, about eight inches deep. The pit had been outlined with rocks, each of which were about six to eight inches in diameter, dozens of them. The grill itself was a thing of beauty, a metal grocery cart turned on its side, wheels removed (I'm guessing those are a fire hazard), the bars bent and flattened in a manner that allowed the meat to cook about six inches above the flames. We could barely make out the words "Piggly Wiggly," sizzling and bubbling, a green (and likely toxic) misshapen blob still attached to the cart in the area where many children must have sat before the cart had been appropriated for such use. Yes, a little creativity, some knowledge from high school shop class, and the willingness to steal a shopping cart for a good cause—and you've got yourself what's known as a redneck barbecue grill. Party on.

I've also learned a thing or two about fancy, creative cocktails, exotic (and homemade) wines, domestic and imported beer, and moonshine during this journey. First, all of it flows freely and is poured generously. Second, sharing is not only OK; it's expected. And third, if you mix up a concoction more than once, it's a secret family recipe. Everybody always seems to have enough for others too, and I learned the hard way after Daytona to politely ask what a drink's ingredients are before downing it. Homemade muscadine and blueberry wine do not mix at all well with a microbrewed beer called Smokey Joe's. As a general rule, I am a wine drinker only. In fact, I am a red wine drinker only. I have always figured that if it's purported to have cardiovascular benefits, I'm actually doing my body a favor by, essentially, juicing.

I let my hair down in 2013, realizing right off the bat that if I were going to really get a taste of tailgating, I had to jump right in and taste everything. I learned early on in the season that I still have a very low tolerance for alcohol, and my tolerance is even lower for mixing different types of alcohol. I had to pick and choose carefully. Basically, if the drink had a cool or whimsical name, I'd try it. There were the legendary drinks called Mississippi Gambler and Alabama Slammer, of course, but there were so many others. I made a point to write down some of the more interesting names and recipes: there was Creek Tea, a simple mix of bourbon, lemonade, and bitters. I sipped a Kissed Caramel Mule in Darlington, an unusual but fabulous mix of Caramel vodka and ginger ale. I toasted with some Louisiana partiers and a Vieux Carre (named for the famous Louisiana hotel), which is a potent blend of whiskey, cognac, vermouth, Benedictine, and again, bitters. Come to think of it, I suppose I should get down on my knees and thank the good Lord that I didn't come away from all of this with an alcohol-chugging monkey on my back. But as they say, when in Rome…

At the more family-friendly tailgating events, there are games set up for families, friends, and neighbors to play: cornhole (another name for "bean bag toss," though I think there are many other names that might have been more suitable), redneck golf (an interesting combination of a short ladder and two rubber balls connected by a string), disk throwing, horseshoes, lawn darts, and yes, there's always at least one beer pong table set up. In short, tailgating is just plain fun, and everybody who's there is there because they love whatever sport is about to be served up: football, baseball, or NASCAR. The biggest difference I saw between the different sports' tailgating is the sheer number of days that NASCAR tailgating lasts. A good race party cranks up sometime on Thursday or Friday, and it winds down late on Sunday, maybe even Monday morning. At some races, namely Daytona, it goes on for an entire week, sometimes longer.

I wish I hadn't felt so intimidated at that first race in Daytona back in February. I was so nervous about being there alone and not having any idea what I was doing that I adopted an "all business" attitude. In hindsight, I regret that. The tailgating was fabulous, with creatively festooned campsites, RVs decked out like I've never seen before or since, and food as far as the eye could see, all of it beautiful and smelling simply divine. In my effort to appear as though I was there on a mission and knew exactly what I was doing, I missed out on the most fun part of any race, which is the party. I did get the chance to talk to two terrific groups of people while I was in Daytona, one of them being the sweet couple who had the misfortune of sitting beside me during the race. The other group, two couples named Barbara and Hap, and Linda and Danny, gave me my first true taste of NASCAR tailgating. What they shared with me about racing was very interesting, and it broadened my perspective of NASCAR in a way that helped me appreciate the depth of the fans' passion. I met them quite by accident while poring over a map of the track, and I was looking lost and frustrated, I'm sure.

Yet another camping setup—infield Atlanta.

Barbara and Linda were walking back to their campsite after doing a bit of shopping (Barbara and Hap are Dale Jr. fans, and Linda and Danny are Jimmy Johnson followers). They took pity on me, no doubt in part because of the casual business summer suit I was wearing, and asked if I needed any help. Any woman knows that that act alone is a rare occurrence. Two women genuinely showing concern for a third, a stranger no less? That would never happen in, say, a mall. I explained to the women as concisely as I could that I was there alone, had never been to a race, and had no idea where to go or what to do before the race actually started. Thank goodness t hey weren't a tag team of serial killers, because with every word I uttered, I exposed my vulnerabilities with reckless abandon, uttering no-nos such as, "I'm here all alone," and "I have no idea what I'm doing." I did my best not to cry, but I am a crier, so I had tears in my eyes. *What was I thinking, coming to a NASCAR race without my husband, without anyone?* I kept asking myself.

Fascinated that there could possibly be anyone left in the free world who had never been to a NASCAR race, the women kindly invited me back to their campsite to spend some time with them and their husbands before the green flag waved. Normally I would never have taken them up on such an offer but somehow, in that weak moment of frustration and, yes, I do believe intimidation, it seemed all right to me. We walked and chatted, and before I knew it, we had reached their RV. I learned later that these two couples, along with every other fan who drove an RV and parked in that area for the race that day, paid about $500 for the spot. Infield camping spots, many of which allow fans to watch the race from the center of the track rather than the grandstands, cost much more than that.

Hap and Danny were busy cooking and watching television when we approached the motor coach. The TV, a giant flat-screen wonder that would have made my husband quiver with joy and speak in tongues, was displayed underneath a canopy. Chairs were casually arranged under the canopy too, and we each moved to take one as the sports announcer narrated that week's qualifying races that had been held in the days that led up to Sunday. Each of the guys at the RV had a beer in hand, and Hap was manning a large pot hanging over an open fire. Inside the pot, water was rolling and boiling, and the mix of spices that rose on the steam hinted at some sort of seafood. It smelled delicious. The two women introduced me to their husbands and explained to them (in a rather scandalized manner) that this was my very first race. Everyone laughed at that thought; I'm not even sure the men believed me.

With the introductions over, the next order of business was fixing me up with something appropriate to drink. It was nearly 11 a.m., and it looked like I had some catching up to do. Danny got to work measuring liquor, slicing fruit, and crushing what looked like brown sugar cubes with a mortar and pestle, while Linda and Barbara showed me around inside the RV.

I consider myself pretty knowledgeable when it comes to decorating and such, but I must say that I walked from one end of that traveling home to the other with my mouth hanging wide open. In a word, the thing was beautiful, and I'm not much for portable homes. It offered up rich wood and leather, a roomy kitchen, and two big bedrooms, and not one but two bathrooms. Custom-made drapes and curtains graced the large windows. The couples had traveled to Daytona from Montgomery, Alabama, and they had done it in style. I doubt that they even knew they were moving along the highway, this thing was so big and cushy. I also noticed the only thing that gave away the fact that the RV belonged to race fans, and that was the sight of competing throw pillows strategically placed on the leather sofa. There were two Dale Jr. pillows, and there were two Jimmie Johnson pillows. They were on opposite ends of the couch. Other than that, we could have been walking through a home featured on one of those decorating shows— "Turn Your Trailer into the Taj Mahal," narrated by Robin Leach. The couples, both retired, own the RV together, and it is used mainly for road trips to NASCAR races. *I could do this,* I thought to myself. *Not only is this not so bad, it actually borders on luxurious. Not at all what I picture when someone says the word "camping" to me.*

Once the tour was over, we went back outside, everyone taking a seat under the protective canopy. Even that early in the day, the sun was warming the east coast of Florida quite nicely. Jimmie Johnson and Dale Earnhardt Jr. pennant flags flapped and chattered busily in the breeze, while the television now featured driver interviews taped days before. Danny handed all of us a heavy, short glass, filled with a brown liquid and garnished with citrus fruit. It smelled sweet, with a hint of maple. What Danny had mixed while we were inside the RV was a delightful cocktail called a Back Forty, which is simply bourbon, citrus fruit, and maple syrup. I think the maple syrup earns it a spot in the "breakfast" beverage category, which was all the go-ahead I needed to start sipping, considering the morning I had already had. As we slowly sipped the delectable cocktails, the two couples began to explain to me just why NASCAR fans love stock car racing so much. They made some very good points, and the things they shared with me that day helped me shape my ideas down the road about NASCAR fan passion.

These two couples met at the season inaugural NASCAR race in Daytona more than twenty years ago, and they've been fast friends ever since. Their children were young back then, and they traveled to every race with their parents through the years. Of course, the RV wasn't around during that time, but the couples still camped. They just did it in tents. Personally, I much prefer the rolling house version

Driver Jimmie Johnson, a heartthrob favorite of many fans, stops and reflects before a race.

of camping to the tent version. Their parents were NASCAR fans too, so the love of the sport has been in both families for three generations now. I have learned since that that is not so unusual.

Even as far back as the early 2013 NASCAR season, I had already had a taste of the openness and the availability of many NASCAR personalities. These couples affirmed what I was already beginning to see. The drivers, crew members, even the team owners have an air about them that says, "We all have the same roots. I am not any different than you are. I just love to drive fast." Fans adore that about their heroes.

As Linda, who had switched from drinking Bloody Marys to the sweet Back Forty, was finishing up a story about how she got

Clint Bowyer's autograph, the two men looked at each other as if on cue and said, "Well, it's probably time," and everyone got up in unison to do a specific job. The guys disappeared into the RV and came out carrying two large coolers. The women went in right after them, then came out carrying round, red plastic trays and rolls of paper towels. Of course, I offered to help with whatever it was they were doing, and of course they declined my offer, since I was a guest.

Hap opened one of the two coolers, and reaching in almost up to his elbows, scooped up two armfuls of red crustaceans (some of them still sluggishly waving their arms as if signaling for a rescue plane) and plopped them into the spicy, rolling water. A few minutes later, he scooped them back out with a large, slotted spoon and dished them onto the red trays, which had been lined with newspapers, then paper towels. He repeated the process several times, until the coolers were empty and he had six or eight trays heaped with steaming crawfish. I had just watched Hap cook about five hundred of the crunchy red creek dwellers, and as he filled the trays, passersby stopped to remark about the bounty, hoping to be asked to partake. Many of the visitors have known these couples for years, so they all knew the drill. It seems that Barbara, Hap, Linda, and Danny have been hauling crawfish to Daytona and cooking it up for themselves and their neighbors for many years now. I suppose you could say that it's become tradition.

February 24, 2013, is the day I learned how to properly eat a crawfish. I had previously resisted doing so for my entire life, although I'm not really sure why. I love shrimp. I love lobster. Perhaps it's the bright red creature's bugged-out black eyes or its long, stringy antennae that bother me. Maybe I just don't like the sounds that people make when they eat them. You see, the proper way to eat a crawfish is to first separate the head from the tail, sucking the juices out of the head in order to get all the delicious Cajun spices that hide in there. I did that part too, just as Hap showed me, fearful the entire time that I'd suck something out of there besides just juice. Do they have brains? How firmly are those little black eyes anchored in there? I tried not to think about it. After you've sucked all the juice and nothing but the juice from the head, you crack the shell that surrounds the meat on the tail and devein the little guy. Then you eat the tail meat. Finally, you suck the claws. Of course, the larger the crawfish, the more meat you'll find in there. The whole process is terribly noisy, improperly messy, and flies in the face of all things mannerly and Southern, except of course that eating crawfish is largely a Southern thing, especially in Texas and Louisiana. The average crawfish lover can eat a hundred or so at a sitting, if he's really hungry. It's a very messy undertaking, and I would advise that you not wear a white linen suit should you ever be invited to a crawfish boil. When I was finally sated, my jacket lapels were stained a pinkish-red, as though I may have witnessed a shooting from afar.

In almost no time, the crawfish were gone. All that remained on the trays were a piece of antenna here, a round black eye there. But while others had helped eat the whole mess, they had also brought food with them to share. We ended up having a delicious feast of crawfish, corn on the cob, big slabs of Texas toast, and boiled potatoes. While everyone ate, we all talked NASCAR—drivers, races, new rules, who stood the best chance of taking it all this year—and while I could contribute very little to the conversation on those topics, I certainly soaked it all in as I feasted on little red crustaceans.

My next tailgating opportunity was at the May race in Talladega, but as I stated earlier, I passed it up. The modest, nutritious, nerdy lunch I packed for the three of us was the extent of our tailgating in Alabama. The hospitality experience, the fabulous pork barbecue, the live band, and the entertaining display to which we were treated by the women very obviously hustling drinks was all enough for me that day.

Also a May race—and my first night race—the race in Darlington, South Carolina, offered up some terrific tailgating. The camping areas were, for one thing, very well organized and clearly marked. The fans were well behaved, even polite, and I had no trepidation about venturing out on my own, wandering around and talking to people, even alone and at night. Many campers had taken great care to decorate their areas with whimsical lighted palm trees, strings of lights, plastic pink flamingos, flowers, and of course, driver pennants and posters. One group of campers had rigged up a big inflatable swimming pool, fitting it snugly into the back of their pickup truck and filling it with water. An oscillating fan set up about six feet from the truck blew a cool breeze over the whole scene. Five people sat in the back of the truck, sporting bikinis and swim trunks, sipping cocktails (complete with colorful paper umbrellas) and beer, and listening to some wistful Jimmy Buffett tune punctuated by steel drums. I had to laugh when I saw the whole getup, and of course my laugh got me invited over for a closer look. This group of fans, all of whom looked to be in their late twenties, had driven to the race from farther south in South Carolina. They were tailgating the easy way, having ordered two six-foot-long submarine sand-wiches to last them two full days. They did, however, have what looked to be a full bar on display just inside their roomy, inviting tent. They shared with me a slice of one of the sandwiches and a beer, and in exchange I left them with a twenty-dollar bill for their trouble. They resisted the gesture, but in the end, common sense was the order of the day. In a pinch, that twenty dollars could buy them a sandwich or two on the way home should the need arise. I have college-age children and as a mother, I would have appreciated the gesture had one of those young people belonged to me.

When the race was over that Saturday in Darlington (that race is one of the few Sprint Cup races held on a Saturday, and I think they do that because of Mother's Day being on Sunday), my friend Frances and I hitched a pricey ride back to the parking lot on a bicycle/rickshaw sort of contraption powered by a young kid who was in great shape. He had to be, pedaling that bike with as many as four passengers back and forth all day long, from the track to the parking lot, half of the trip being an uphill one. Exhausted, Frances and I were not really paying attention when we told him where to drop us off. We were still quite a distance from our car and didn't realize it until Mr. Hardbody had pedaled off to overcharge another unsuspecting group of passengers. We ended up walking a pretty long way in the dark, through a camping area, trying to make our way back to the car and to our husbands. Not once did I feel unsafe, nor did I need to. Everyone seemed to be minding his own business. We did not encounter anyone being rowdy or behaving as though they were excessively intoxicated. We of course made our way back to the car and to our waiting husbands, who had beat us back to the car with a good half-hour to spare, rickshaw notwithstanding.

I have to add that the hospitality at Darlington, and I'm sure at other races, is not confined to the camping areas. We had great seats at this race in South Carolina, surrounded by fun, not obnoxious, people. For instance, every time Jimmie Johnson took the lead at that race, someone sitting just above and behind us would pour and distribute shots of tequila. While I personally did not indulge in those little mini-celebrations (tequila is my archenemy of many years, going all the way back to my college days) a lot of fans sitting around us did. Pretty soon, everyone sitting in our section was a Jimmie Johnson convert, simply because the payoff was so much fun. We had an entire section that, by the end of the race, had become a devoted Jimmie Johnson fan club. Yes, I can see how lifelong friendships are formed and nurtured in the welcoming environment of NASCAR tailgating.

At this point in the season came an extended period—a couple of months, to be exact—in which I did not attend any races in person. NASCAR tracks in the northeast, midwest, and west hosted many of the summer races, and by the time I got the opportunity to go to my next race here in the South, I was primed and ready. I had sorely missed the smells, the people, and, oh, had I missed the sounds. It seemed quite fitting, then, that the next race I attended was at Bristol Motor Speedway, perhaps one of NASCAR's most notorious, most legendary, racetracks.

Chapter Seventeen
Thunder Valley

Loud. If I had to sum up the weekend I spent in Bristol, Tennessee, using just one word, that would be it. Loud. People had warned me about the noise level at Bristol Motor Speedway races, but warnings didn't do the phenomenon justice. Nothing I had devoured in my reading did it justice. A short track surrounded by the steepest grandstands I have ever seen in any sport, Bristol Motor Speedway is a lot like a concrete megaphone. I could actually feel my internal organs vibrate when the powerful cars started up at the Friday and Saturday night races, and if I'm not mistaken, my hearing is not quite what it was before going to that fabled track in Tennessee in late August. Without the insulated cushioning that my headphones provided, the noise was quite disorienting. Now I understand completely why Bristol Motor Speedway has the nickname "Thunder Valley."

This half-mile concrete oval situated very near the dividing line between Tennessee and Virginia is one of NASCAR's most popular tracks. Dubbed "the world's fastest half-mile," at 0.533 miles long it's not the longest track in the Sprint Cup Series, nor is it the shortest (that distinction now belongs to the 0.526-mile-long Martinsville track in Virginia). It's not the fastest track either, not by a long shot. No, fans tell me that it's the concrete track, two pit roads, and the dizzyingly high banking that make Bristol such a hotspot. This track also holds the NASCAR record for the most caution flags waved during a race, as contact between cars is pretty much inevitable on such a short road.

Bristol Motor Speedway is the fourth-largest sports venue in the United States and the eighth-largest in the world. I also heard another interesting fact that one of the television race announcers threw out for consideration in prerace banter: Bristol is a small city as cities go, with a population of about thirty thousand people. When NASCAR races come to town two times a year though, Bristol becomes the third-most populated city in Tennessee, right behind Memphis and Nashville. For a long time, Bristol's second race of the season, which is the night race we attended in August, was known as "the toughest ticket in NASCAR" to obtain. There were years-long waiting lists for both tickets and camping spots. Today, that is no longer the case because again, today's unfriendly economy makes it harder for fans, especially those with families in tow, to afford making a weekend of this important NASCAR race.

This particular track seats about 160,000 people. Those people are seated in a stadium setting that looks very much like a soup bowl (thus another of the track's nicknames, "The Soup Bowl"), because the track and grandstands are banked so steeply. As many drivers have said over the years about this track in the northeast corner of Tennessee, the sharp banking "makes for a lot of paint swapping." So while the speeds on such a short track might not be the fastest in NASCAR, the thirty-degree banking in spots makes it very tricky for drivers to navigate without bumping and scraping one another. Because of the difficulties that are naturally going to occur with forty-three cars driving very fast in a tight, high-banked oval, I knew enough about NASCAR races at this point in my journey to know that I was in for a real treat.

Bristol Motor Speedway opened in 1961 and was built by Charlotte Motor Speedway enthusiasts Carl Moore, Larry Carrier, and R. G. Pope. These three gentlemen were inspired by the track they saw in Charlotte the year before, which had been built by Bruton Smith and Curtis Turner. The three decided to build a shorter,

When drivers race neck-and-neck at speeds exceeding 100 mph, paint gets swapped.

half-mile track with more intimate seating and requiring less real estate than its predecessors. The developers scratched many of their design ideas on paper bags and napkins and after about one year, Bristol Motor Speedway was complete and ready to host its first race. Back then, the grandstands seated about eighteen thousand people. The parking lot held twelve thousand cars. The track itself was a perfect half-mile, measuring sixty feet wide on the straightaways, seventy-five feet wide in the turns, which were then banked at a relatively tame twenty-two degrees. The final cost for both the land and the construction was a whopping six hundred thousand dollars.

Driver Fred Lorenzen won the first pole position at Bristol back then with a searing 79.225 mph practice time. The legendary Rex White drove in that inaugural race named the Volunteer 500, and so did fan-favorite David Pearson. Country music legend Brenda Lee, then only seventeen years old, sang the national anthem.

When the Volunteer 500 was over in 1961, Jack Smith of Spartanburg, South Carolina, had won it. But as one might guess in NASCAR, there was a story that went along with that win. Smith wasn't actually in the driver's seat when his Pontiac crossed the finish line. He drove the first 290 laps, but by that time the tremendous heat coming from the floorboard had blistered his feet so badly that he couldn't finish the race. Johnny Allen from Atlanta took the wheel as Smith's relief driver, maintaining the lead and ultimately crossing the finish line first. Fireball Roberts, Ned Jarrett, Richard Petty, and Buddy Baker rounded out the top-five finishers that year. Unfortunately, the same scenario that Smith and Allen played out would become commonplace at Bristol races, until power steering and better insulation in the cars were introduced, cooling the floorboards somewhat and taking it a little easier on the drivers' feet. The total purse for that first race at Bristol Motor Speedway was $16,625, the winning driver's share a staggering $3,225. Of the forty-two cars that started the race, only nineteen finished the grueling competition. Bristol soon became known among drivers as "the race to win," where skill and endurance—not merely luck and having a superior car—separated the good drivers from the great ones. For eight years, the track remained just as it had been built, but then the owners decided to shake things up a bit.

In 1969, BMS owners Carrier and Moore decided to have the entire track excavated and reconstructed, this time building much steeper banking into the turns. The sport was abuzz about the changes, and when the first advertisement launched for the new track, fans and drivers alike read about thirty-six-degree banks in the turns, a challenge that was unheard of up to that point. The banking was so steep that it added to the length of the track, changing it from a half-mile to 0.533 mile. Still, the track is so short that at the start, the cars in the back of the pack (those who qualified at slower speeds) are already about a half a lap down before they've even moved. That many cars in that tight a space racing at speeds over 100 mph can mean a lot of caution flags, dust-ups, and wrecks, even though the average speed is much slower than at longer tracks.

Compared with the superspeedways of Daytona and Talladega, short track races require entirely different racing strategies. It never occurred to me before I started studying NASCAR that the length and banking of a track might have any impact on the race itself, but they certainly do.

The length of the track and the sharply banked turns were not the only things that changed with the 1969 renovation. Right away, drivers saw differences in the feel and speed of the track. While the initial track record at BMS was Lorenzen's 79.225 mph in 1961, after the track was changed, the official record was the 88.669 mph lap that Bobby Isaac used to earn the pole position for the March 1969 Southeastern 500. Isaac's speed was more than 9 mph faster than Lorenzen's. In July of that same year, at the Volunteer 500, Cale Yarborough shattered the record by nearly 15 mph with a pole-winning speed of 103.432 mph. Think about that for a minute. Yarborough's car traveled a half-mile track at a speed of more than 100 mph. I get motion sickness just thinking about that. It reminds me of a ride we used to enjoy as kids when we went to Six Flags. Riders would all have to stand straight and tall against the walls of a round room. Once everyone was in and situated, the room would begin to spin, and it would spin faster and faster until it felt like your head weighed about a hundred pounds, and you could not lift it off the wall to save your life. Once the spinning hit a certain speed, the floor would drop out from under your feet. Of course, there was always at least one kid in the room who thought it would be funny to spit while spinning at dizzying speeds, so of course he would. Invariably, his spit would hit three or four people in the face—hard—before it eventually evaporated or just hit the wall. Wonder if Cale ever spit inside his car when he was driving that fast? David Pearson won the first race on the new track, leading about three-quarters of the five hundred laps and beating Isaac to the finish line by three full laps. Despite the excitement and big-name stars that Bristol Motor Speedway generated and attracted, the owners had a hard time keeping the facility in the black, and in 1977 they sold it to Lanny Hester and Gary Baker. In spring 1978, the speedway was unveiled and renamed Bristol International Speedway. In April, Darrell Waltrip won the first of his career twelve wins at Bristol. Hester and Baker saw early on that the overwhelming heat of the August Volunteer 500 race was likely a factor in lagging ticket sales for that race (I could have told them that); therefore, the first night race in Bristol was run on August 26, 1978. Many legendary NASCAR drivers have made history at Bristol, and even though track ownership has changed hands a few more times since that 1976 sale, Bristol remains one of the fans' favorite venues. History has been made at this brutally tough track for decades. On April 1, 1979, a little-known driver named Dale Earnhardt earned his first major Cup win in a #2 Chevy. Dale Jr., Darrell Waltrip, Rusty Wallace, Bill Elliott, Alan Kulwicki—all drivers whose legends persevere over the decades—have left a mark on Bristol, or perhaps Bristol has left its mark on them.

In 1992, the last race run on asphalt at Bristol, owner Carrier decided to make the race surface concrete instead of the traditional asphalt. Better tire traction had evolved over the years, and higher speeds had him patching the track or applying an entire new coat of asphalt every few years. Operating costs were soaring as a result. In August 1992, the first NASCAR Cup event was run on concrete in Bristol. In 1996, NASCAR heavyweight Bruton Smith bought the track from Carrier for a cool $26 million and in May 1996, the name was changed back to Bristol Motor Speedway. All through the years since the track was originally built, seats were being added. In August 1998, the seating capacity was about 131,000, with an additional one hundred skyboxes. By the year 2000, seating capacity was increased to an amazing 147,000, when construction of the Kulwicki Tower and Terrace was complete (In 1993, Alan Kulwicki and three others died in a plane crash as they traveled to Bristol for the April race). An infield pedestrian tunnel was added. A new backstretch grandstand was built in 2002, and even more luxury suites were added. Smith has made about $50 million worth of improvements to the facility since he bought it.

Bristol is famous for dramatic races and dramatic wins, one of the most famous of which in 1999 saw Dale Earnhardt and Terry Labonte battling for the lead on the final lap. Earnhardt tapped Labonte and sent him spinning, and while Labonte spun out of control with his tires spewing black smoke, Earnhardt crossed the finish line and made a trip to Victory Lane for the ninth and last time. That race and that finish have been voted "most memorable" by fans many times since then.

Darrell Waltrip is the all-time winningest car driver ever to race at Bristol (twelve wins), and Junior Johnson is the car owner who boasts the most wins (twenty-one wins).

When I was planning my trip to Bristol, I did it with the knowledge that my husband would be staying home while I hit the road with a girlfriend to see the race. My friend's cousin, who is both a model and an avid NASCAR fan, lives near Bristol and never misses a race there. The "girls" in our girls' weekend, then, consisted of a former model, a hairdresser, and a writer—not what most would call a typical gaggle of NASCAR fans. In fact, we were such an unusual trio that we seemed to be a bit of an oddity to most folks at the race, but I will say that everyone was very friendly and most gracious; I can't think of one jovial tailgating group that we were not invited to join, and many fans also visited our well-decorated little canopy, which was tastefully bedecked with fresh flowers and strings of checkered-flag lights. Our tailgating area was obviously put together by three women. In all of the tailgating encounters I had this season, ours at Bristol was the only one with hummus and Skinnygirl mojitos on the menu. We got a lot of funny looks and questions like, "What the heck is that, anyway?" but still, ours was a popular stop on what I've come to call the tailgate stroll.

Late August in the South is a miserable time of year, at least as far as the weather is concerned. Wet heat has laid stagnant over this part of the country for a few months by then, and even a short walk from the front door to the mailbox causes a person to break out in a shiny sweat. Cicadas, those relentless summertime gossips whose concertos reach a frenzied, almost crazed peak in August, buzz back and forth to each other throughout the long days, their calls somehow emphasizing the suffocating heat. I always imagined as a kid that their insistent pleas might somehow convince fall to show mercy, come early, and bring some cool, dry relief. In August, those of us in the Deep South still have two or three more months of sweltering heat to endure before cooler, drier days pay their first visit. August is, in fact, the dog days of summer.

Two women pack for travel and outdoor activities in such weather in a manner far different than packing for any other time of year or type of activities. I knew by the time I was preparing for this race that these events are all-day activities, and going to both the Nationwide and Sprint series races on Friday and Saturday nights meant two long days and nights in the heat and humidity. Hair styling tools and products, therefore, would be pointless to pack. Instead, we opted for baseball caps, scarves, headbands, and visors. Makeup would be kept to a minimum; in the heat, it all slides off after about an hour, anyway. Clothes would have to be loose and lightweight, because even though the August races in Bristol are run at night, the heat never ebbs. By late summer, the air has baked for so long that it just stays hot, hanging over both night and day without much of a variation in sticky temperature. On the upside, I wasn't too worried about makeup, clothes, and such by the time the NASCAR season had progressed to late August. I was more interested in the cars, the people, and the inevitable show that was surely on tap in Bristol just as there had been at every other venue I had visited up to that point. I have to be truthful and admit that even by August, I paid less attention to the point standings as I did such things as the NASCAR rumor mill and the cars themselves. It's a juicy, captivating sport for so many different reasons. At times, the backstories in NASCAR can steam like soap operas, and they are often played out right on the track during a race.

Had it not been for the protests of a few neighbors living in nearby Piney Flats, Tennessee, back in 1960, the Bristol track would have been built seven miles farther south and called "Piney Flats Speedway," a name that I think would have been just awful. In my opinion, things worked out for the best, because I think "Bristol" sounds much better. I read somewhere that racing more than forty powerful stock cars at Bristol Motor Speedway is a lot like flying fighter jets in a gymnasium might be. Fans with whom I talked while we tailgated said it's a lot like watching forty-three cars on the spin cycle in a washing machine. I'd say they're both accurate descriptions of what a race at Bristol Motor Speedway looks like when it's under way.

I had also learned enough about attending races by August to know that I could and should specify where we wanted to sit at the track. The location of the seats greatly impacts the fan experience. I learned at Daytona, for instance, that I prefer sitting high up in the grandstands so that I can see the entire track at all times. I learned at Talladega that sitting on the first row is just a loud, messy, frustrating experience, and the fans do not pay attention to the signs instructing them to sit down. At least that was our experience. At Darlington, I learned that having a clear line of sight to pit road is an absolute must; watching those lightning-fast teams do their very precise thing in a flat twelve or so seconds really is fascinating and adds to the excitement of the race. I had read and heard so much about the "Bristol Experience" that I wanted to be a savvy ticket buyer when the time came. There are actually several good seating options at Bristol, depending on what type of race experience you enjoy. If you select seats along Thunder Alley (near the drag racing strip), the cars run so close together that your seat shakes constantly, and the view of pit road is terrific. In the area known as Wallace Tower, the entire view of both the track and the beautiful mountains that surround it is breathtaking. The Pit Road Party Zone is a completely different experience, with a view of the race from the infield grandstands, access to the pits (which is always fascinating to me), and a marvelous catered meal. As tempted as I was to choose this last option, I reminded myself that I was saving the infield experience for the big Labor Day race at Atlanta Motor Speedway and that I would be leaving for that race in less than a week. I've had that weekend circled on my calendar practically since the beginning of this whole journey, when I had the pleasure of interviewing Ed Clark, and he so graciously offered to find us a spot in the track infield and squeeze us in. I'd hold out for Atlanta to experience a NASCAR race from the other side of the tracks. I had waited this long; another week wouldn't kill me.

We ended up choosing seats in the Wallace Tower section of the grandstands, and I'm glad we did. Whatever faint puffs of breeze that found their way into the stadium that night could be felt up there and were delightfully welcome. Even in the mountains of Tennessee, a Southern summer is not for sissies.

By the time August rolled around in the 2013 NASCAR season, Clint Bowyer was situated firmly in second place, points-wise, among the drivers, and that was good news for me.

The not-so-good news in August was that Bowyer trailed Jimmie Johnson, who had a stronghold on the top spot, by a whopping 49 points. For a good chunk of the 2013 season, it seemed that no one could come close to Johnson in the points race. Many fans had told me that they were getting sick of seeing #48 up front all the time, but whether you're a Johnson fan or not, you have to admit that he's a good driver with a great team behind him. He's also very cute, but I think he is well aware of that fact.

Bowyer came in fourteenth place on Saturday night at Bristol, which may not sound great, but he clinched a spot in the Chase for the Sprint Cup (what I call the NASCAR "playoffs," because that's what they are, really) with that finish. Driver Matt Kenseth won the Saturday race in Bristol, making it his fifth season win and securing him and his #20 Toyota a wild card spot in the Chase for the Sprint Cup. Driver Kasey Kahne threatened to overtake Kenseth in the last laps of the competition, but Kenseth limped to victory on a nearly empty fuel tank. I liked what Kahne had to say about the finish; paraphrasing, he said that he is not a driver who will intentionally wreck another one just to take first place. I have learned over these months that many drivers will do just that, and I still can't get used to it. A NASCAR wreck is really something to see, and when one happens I still hold my breath and close my eyes until the dust has settled and the announcer assures the crowd that everyone's OK.

Hank Williams Jr. headlined the Irwin Tools Night Race pre-race entertainment on Saturday night at The Last Great Coliseum (yet another nickname for the Bristol track), and who doesn't get fired up listening to him? Even I can dance to his music, and I'm not much of a country music fan. The crowd was already in a rowdy good mood well before the race began on Saturday, and the patriotic festivities leading up to the start of the race just added to the mushrooming enthusiasm. Saturday night's race was of course the big event of the weekend, and the powers that be at Bristol not only gave a respectful nod to the nation's military at that race, but they also saluted the "working man" with the Irwin Tools sponsorship. The fans ate it up; by race time, it didn't take much to whip them up into a roaring multitude, eager to see some serious racing and enthusiastic paint-swapping on the track.

The people sitting behind us in the stands had thoughtfully and neatly packed a cooler of red, white, and blue Jell-O shots (I'm telling you, NASCAR is a patriotic sport), and they graciously offered my friends and me a couple as the excitement mounted. Truth be told, I didn't even know what the wiggly, colorful things were when I first saw them. I thought they were tiny, personal-size desserts. My friends and I declined our neighbors' generous offer to slurp down a shot with them right before the engines started. That small but persistent internal voice that only mothers can hear whispered to me even above the deafening din: *Don't do it. Never accept a (drink/Jell-O shot) from a complete stranger. It's dangerous! It's poor judgment! It's trouble!*

I don't know what came over me, but just a few minutes after I politely but resolutely said no to the offer, I actually accepted not one shot but two, something I have lectured my own children about often over the years. What was I thinking? Both of my girlfriends looked at me with such shock and surprise that I had to laugh—and throw back a third shot. When I did that, my friends joined me in my reverie, and that made our generous neighbors very happy. Such nice people. My only defense for my recklessness is that I had gotten carried away by the swell of excitement, patriotism, and the celebration of the blue-collar work ethic. I just couldn't contain my excitement.

In addition to downing the first colorful Jell-O shot I have ever ingested, the events immediately preceding this race had a few added surprises, some of which I had not experienced yet. First, American boxing ring announcer Michael Buffer, with his famous, "Let's Get Rrrrready to Rrrrrrumble" phrase recognized around the world, performed the driver introductions. Fans got the opportunity to

vote (via text, of course) for which driver's introduction and associated song they thought was the best. From a marketing standpoint,
this was a clever move, making the fan experience interactive—
another nod from NASCAR to future, younger fans.

The children of the drivers sang an endearing and rather
moving rendition of the National Anthem, and a ten-plane flyover drew cheers and a stadium-wide "wave" from the estimated
one-hundred-thousand-plus crowd. Buffer's rich, experienced
announcer's voice also signaled the drivers to start their engines,
and while the crowd cheered, that familiar aggressive rumble and
smell of oil and exhaust fumes rolled up and over the grandstands
as the slick and powerful engines obeyed the announcer's command with a lunging, gasoline-fueled growl. Hearing that sound
still makes me weak in the knees. I can't explain it unless I use
my daughter's terminology for something or someone she thinks is
sexy—the cars are just plain "hot." Even as high in the stands as we
were sitting, the floor beneath our feet actually did rumble, and for
a moment I was reminded of an earthquake I had experienced years
ago while on business in southern California. The effect, again, was
a little disorienting, but by that time, I was so swept up in the excitement that I wouldn't have cared if it had been an actual earthquake.
If it had been, I would have just sucked down another wiggly shot
and continued to enjoy myself. Yes, that Saturday night in August
redefined the word "loud" for me, probably for the rest of my life.

In the end, driver Matt Kenseth edged out clean-racing nice
guy Kasey Kahne for the Sprint Cup race win that Saturday night.
Though Lord knows Kahne arguably had every reason to intentionally wreck Kenseth (Joe Gibbs racing team drivers, which include
Kenseth, have wrecked Kahne several times this season), he didn't
do it. I like that about Kahne. So many drivers have reputations for
being nasty, vindictive, and grudge-bearing, but Kahne remains one
of the few truly good guys in the sport.

My friends and I left that race late Saturday night sweaty and
exhausted. My girlfriends actually bought driver T-shirts of their
own; I stubbornly stuck with my Clint Bowyer garb—denim shorts
and a black tank top with silver lettering that boldly proclaimed

me to be a "Bowyer Racing Diva." I looked like the stereotypical "redneck" woman, a sweaty race fan wearing black racing garb, and a tank top to boot. Oh, my mother must be turning in her grave, simply mortified that her heretofore properly behaved, well-dressed daughter had sunk to such a "lowly" place. Sorry, Mom, but I'm having fun.

In less than a week, I'd be headed to Atlanta Motor Speedway, and then I'd really understand what it is to fully enjoy NASCAR's particular brand of fun.

Chapter Eighteen

The Biggest Labor Day Party in the United States

With barely enough time to drive back from Bristol and wash all the laundry that I had brought home with me, it was time to pack up and leave for the big race weekend at Atlanta Motor Speedway. Finally, the event for which I had waited practically all year was here. The Atlanta race would be the first one at which we'd be camping in the infield area of the track.

Atlanta Motor Speedway, originally dubbed Atlanta International Raceway, was built in 1960. Its dubious debut sputtered a bit because of financial speed bumps and changes in ownership. When the first race was held there in July of that same year, famed *Atlanta Journal* editor and sports columnist Furman Bisher stated emphatically that the 1.5-mile track wasn't ready to be used. His assessment of the Raceway included the fact that there was mud everywhere, fans in many of the front-row seats were sitting too low to see over the retaining wall, and the "facilities" consisted of a three-hole outhouse. That depiction of the racetrack is a far cry from the majestic structure that is today called Atlanta Motor Speedway, a sprawling track situated on 887 acres just south of Atlanta in Hampton, Georgia. But in the early days, the fledgling facility struggled, and in the 1970s, it was financially reorganized under Chapter Ten bankruptcy. The track subsequently came under the direction of several general managers before finally settling down under the watchful eye of the late Walt Nix, who served in that capacity for almost twenty years. Ed Clark has been at the helm for more than twenty years.

In spite of the difficult years of financial struggle, the Atlanta track had still flirted with and attracted the attention of several local celebrities. Portions of *Smokey and the Bandit II* and *Stroker Ace* were filmed there.

When Jimmy Carter, an avid NASCAR fan and former AMS ticket vendor in the 1960s, was running for the office of governor of Georgia, he promised a barbecue dinner at the governor's mansion if he won the election. He did win, and he made good on his promise to all his racing cronies. Naturally, the racing community enjoyed the subsequent barbecue dinner Carter threw to celebrate his 1978 political victory—at 1600 Pennsylvania Avenue in Washington, DC.

The struggling racetrack in those early days consisted of the Weaver Grandstand and some uncomfortable wooden bleachers. Many race fans would just bring blankets and sit on the dirt banks surrounding the track to watch races. All that changed, however, when NASCAR heavyweight and Speedway Motorsports owner Bruton Smith bought Atlanta International Raceway in October 1990. Right away, he renamed the track Atlanta Motor Speedway. A year after that, the East Turn Grandstand was built, increasing the seating capacity by twenty-five thousand. Smith also added thirty enclosed suites along the top of the open seating, and those suites were more luxurious than any race-goers had seen up to that point.

Expansion continued at AMS, and the now thriving track began hosting such races as the Nationwide Series, ARCA (Automobile Racing Club of America), drag races, and Indy car racing. Concerts, business conventions, and even dog shows have been hosted by the community-friendly Atlanta Motor Speedway.

In the mid-1990s, Tara Place was built, a beautiful nine-story building that literally leans out over the racetrack and houses forty-six luxury condominiums, a swimming pool, and tennis courts, as well as the AMS corporate offices. Growth still continued with construction of the Tara Clubhouse, the expansive Earnhardt

Grandstand, and then the Champions Grandstand. The number of luxury suites grew to an impressive 137.

Interestingly, when the Champions Grandstand was added, the start/finish line was relocated from the west side of the track to the east. Two doglegs were built into the front stretch, making the track a 1.54-mile quad-oval and one of the fastest on the NASCAR circuit. New media facilities, garages, and many fan support buildings were also added, making Atlanta Motor Speedway a state-of-the-art sports sanctuary.

October 2006 saw yet another structure—the Winners Grandstand—added to the sprawling facility, giving fans an unparalleled view of the front stretch and pit road. At just about the same time, trackside luxury RV parking replaced the aging Weaver Grandstands. I've seen that parking area, and I can tell you that in all honesty, I believe I could live there year-round in one of those gorgeous RVs in one of those parking spots and be perfectly happy. When NASCAR says "luxury," they mean it. Adding to the plush amenities, Smith then paid to have Club One built, a public suite limited to just one thousand fans. Club One is climate-controlled (in Georgia in the summer, that's a must), offers a fabulous view of the track, and features a rooftop observation deck.

In early July 2005, Tropical Storm Cindy plowed through the southeastern United States and spawned a wicked tornado that headed straight for Hampton and Atlanta Motor Speedway. The tornado cut a destructive half-mile-wide path and stayed on the ground for an incredible four miles, splintering structures, trees, and even a couple of airplanes in its path. Ed Clark was president of AMS at the time, and he remembers the damage and destruction all too well. "Everything except the track itself was damaged to some extent," he said. Grand old oak trees, flagpoles, and light poles were snapped in half like twigs, and the rooftops of many of the condominiums were caved in. Entire sections of the grandstands were chewed up and spit out by the twister. Ironically, the bowl-shaped arena that is Atlanta Motor Speedway helped the tornado form and take its devastating shape, giving it enough muscle and definition to wreak havoc at the Speedway, then go on to do further damage at nearby Tara Field Airport. When the category F-2 tornado with its 157 mph winds stormed out of the Speedway, it left behind about $40 million in damage. Unbelievably, although the Speedway's events for the next couple of weeks were cancelled, that year's October 30 NASCAR Bass Pro Shops MBNA 500 race was run as planned. Construction, repair crews, and the AMS staff worked day and night to make that happen.

Although Atlanta Motor Speedway is known for being one of the fastest tracks on the circuit, the facility now hosts just one race weekend a year these days. In 2009, that one race was changed from an October date to Labor Day weekend. Not so long ago, there was also a spring race scheduled every year, but that race was removed from the regular season schedule. A Sprint Cup race was in turn "given" to the Speedway Motorsports track in Kentucky. Many devoted fans of Atlanta races were devastated by the move, and metropolitan Atlanta businesses—especially those located near the track in Henry County—lamented the loss of millions of dollars in revenue that AMS race weekends bring to the area.

Race promoters cited Atlanta's unpredictable spring weather and declining attendance as reasons for the move. I really can't argue with that point; in an Atlanta springtime, one is as likely to see a scorching-hot day as several inches of snow. In an *Atlanta Journal-Constitution* article about the decision that impacted so many fans and businesses in Georgia, Clark was quoted as saying, "This is business. Our company looked at it like we can do one event in Kentucky and make it huge and do one event in Atlanta and make it huge, and the overall gain to the company is more than doing two races in Atlanta." In fact, reverberations caused by the decision to yank the spring race from the Hampton track carried all the way to the gold dome of the state capitol. An alarmed Governor Sonny Perdue, Lt. Governor Casey Cagle, and House Speaker David Ralston, fearing the loss of the estimated $90 million NASCAR race weekends bring to the area as far north as Atlanta, rushed to NASCAR headquarters in Charlotte to plead the state's case and try to reverse the decision. Their efforts failed. Still, the Labor Day weekend at Atlanta Motor Speedway remains a huge boon to the local economy and a beloved annual tradition to about one hundred thousand fans every year. Throughout much of the rest of the year, Friday Night Drag Races, Thursday Thunder races, and other events such as 5K foot races, concerts, Zombie Runs, and more keep the venue hopping and the community coming back for more.

Every track on the NASCAR circuit offers infield camping for fans, but the Labor Day Weekend race in Atlanta was the one I wanted to experience for myself. When Ed Clark graciously offered to find us a spot to do just that, I couldn't believe my good fortune. Those spots are usually reserved years in advance; his offer was akin to striking racing gold, especially for a newbie like myself.

Of course, as may be evident by now, I'm not much of a camper; therefore, my husband and I do not own a camper. Make no mistake though, we'd have gone out and rented one for the weekend if necessary. There was no way I was going to let this opportunity slip through my fingers. As a journalist, there is no better way to relate an experience than to live it. With all the races I'd been to in the 2013 season, with all the traveling and interviews and tailgating experiences I'd logged since February, I had yet to do just this one thing. Because I hadn't experienced infield camping yet, I still felt like a racing outsider. It seemed to me that every respectable NASCAR fan had at least one treasured infield camping story to share. After Labor Day weekend, so would I.

Steve and Frances, friends of ours for years and NASCAR lovers from way back, have a camper. They've used it for both family beach vacations and for NASCAR races. When we asked them if they'd like to go to the Atlanta race with us and camp in the infield, they were thrilled. Orthodox NASCAR fans who have camped at races for years, including both Bristol and Atlanta, they still had never had the opportunity to camp in the middle of all the action. It was a win-win for all of us.

Camping with these two would be like camping with walking, talking NASCAR Wiki-people. I am continually amazed by the knowledge that they share between them with respect to NASCAR. Whether it's horsepower, rules, tire technology, or who's dating whom on the NASCAR scene, they know the answer.

Now Steve and Frances, longtime season ticket holders, had their own camping spot for the Bristol August race for many years; as Frances explained to me, there was a time that claiming such a spot at Bristol was like possessing a precious family heirloom. That spot, and the season tickets to that same race, were gems to be treasured and never let go. When they had camped at AMS in the past, they had done so outside the track very near an old, lone graveyard that is visible from the top of Turn One. Watching the race from the center of all the action, rubbing elbows with thousands of race fans, and getting up close and personal with drivers, mechanics, and those amazing cars would be an entirely different experience even for them. Labor Day weekend 2013 promised to be an exciting first for all four of us.

Strategizing is critical when camping; prerace planning was essential. We'd be at the racetrack for four nights, Thursday through Sunday, and we had to plan accordingly. There would be no wandering out in the evenings to find fun, out-of-the-way restaurants. Once that camper is parked and set up, that's where it stays. Besides, it's not really camping if you're restaurant-hopping, now is it?

Given Frances' camping expertise, I followed her lead. Between us, we planned out every meal in advance and brought along everything necessary to prepare those meals. Their camper is a very comfortable one, with two bedrooms, a full bathroom, and a nicely equipped kitchen. Oh, and it's air conditioned. I think that was my favorite thing about this very nice, full-service camper. The air conditioner worked like a charm all through that stifling, ninety-degree-plus weekend.

Planning and preparing for the meals was the easy part; packing clothes for this race proved to be a bit more of a challenge for me. I knew I'd be talking with people whom I'd met before: Mr. Clark, his wonderfully efficient executive assistant Juanita, the helpful and accommodating credentials people, and some key people with whom I had set up interviews. I wanted to dress like a professional for those encounters. We may have been camping, but that didn't mean I had to look as though I was camping. Therefore, I packed several dresses, some jackets, and other such articles of clothing for those meetings. I packed all of my makeup and toiletries, including hair products and my favorite summertime perfume. I even brought along two of my best summer nightgowns. My husband sweetly obliged my overpacking and the two large suitcases, bulging at the zippers. When Frances saw what I had packed, all of which could have easily gotten me through two weeks in Europe instead of a long weekend at a racetrack, she just looked at me as though I had lost my mind, and smiled. An old pro at this sort of thing, she had efficiently packed shorts, T-shirts, baseball caps, and walking shoes. She also packed everything she'd need to snag some great auto-graphs: namely, twenty Sharpie markers and a brand new NASCAR hat (her dog had eaten her old one). Frances was a woman on a mission that weekend.

Because of work schedules and other commitments, we arrived in two groups. Steve took the camper down to the track on Thursday, when the gates opened to campers. Frances, my husband, and I would drive down to Hampton on Friday at around 5:00 a.m. Frances was determined to see the hauler parade, an event that takes place before every race and unofficially kicks off the weekend's festivities. The enormous trucks, having hauled their precious cargo to Atlanta from Bristol, would be lined up along the road that leads to the Speedway in the wee hours of the morning. On cue, they would enter the stadium through the tunnel one at a time, a slow, lumbering procession that I thought was best described by Juanita, Ed Clark's assistant: "I love the hauler parade. It's the only thing NASCAR does that's slow; it's like their version of a ballet."

I'll admit that I would have been just fine having that description of the hauler parade, rather than having to get up at 4:00 a.m. to see it for myself. Frances insisted though—she reminded me of a kid on Christmas Eve. It was kind of cool to watch the sun rise while those gigantic trucks, their trailers plastered with larger-than-life photos of Danica Patrick, Dale Earnhardt Jr., Jimmie Johnson, and all the others, creep onto the track entrance and through the tunnel, then ease along the short road that led to the spot where they'd be spending the next few days. It was also a bit sobering to know that inside each of those trucks slept two powerful cars, both of them meticulously engineered and costing a small fortune. One of those cars would be awakened very soon, allowed to stretch and warm up on the practice track, compete against other such machines in qualifying, and then be let loose, wide open, in an attempt to win a very important race on either Saturday or Sunday night. You see, the Labor Day race in Atlanta is the next-to-last race of the regular season. The Richmond race would be held the following weekend, and then the top twelve drivers emerging from the regular season would begin the Chase for the Cup in Chicago the weekend after that. The Chase stirs as much excitement as do playoffs in any other sport, and winning often comes down to the very last laps of the very last race.

Our friends knew that my husband and I would need some coaching before being turned loose to explore inside the track at Atlanta, so they sat us down and explained a few things. Friday would be the day to spend as much time as we possibly could in the garages, because the crowds would only get bigger as the weekend progressed. Our credentials, also graciously arranged by Ed Clark, would give us wonderfully close-up access to the cars, the garages, the mechanics, and the prerace inspection process—everything I had read and heard so much about (but had only glimpsed from a distance) up to that point. I'm not much of an autograph collector, but Frances patiently explained proper autograph-collecting etiquette to me anyway. Apparently things can get nasty when a starry-eyed fan gets a little too eager to obtain his favorite driver's John Hancock.

The two of them were careful to explain that although our credentials did gain us access to the garages and pit area, there were still some actions that would simply not be considered acceptable. For instance, while it's perfectly suitable to look inside a garage and watch the mechanics work, it is absolutely forbidden to just stroll inside the garage and start poking around and asking questions, or heaven forbid, to start touching things. The issue is one of safety, of course, but there is also a level of secrecy and proprietary preparation that teams simply do not want to share with just anyone. I saw a few overzealous fans rebuked that weekend for just such behavior.

Once all the ground rules were explained, it was time to venture out and see what we could see. Friday, then, consisted of an early rise, the hauler parade, and watching the teams and officials prepare for the Saturday and Sunday races.

I had called Kyle, the ever-so-helpful mechanic from Penske Racing, to see whether he'd be at the race but unfortunately, he would not. He did tell me to be sure Saturday to stop by the #12 hauler (the truck that had transported driver Sam Hornish's cars, tools, and parts to the race) and ask for someone named "Pickle." Kyle said that Pickle would take good care of us. Eager for the opportunity to hopefully tour the inside of one of those gigantic garages-on-wheels, I did exactly as Kyle had instructed.

Pickle, as it happens, is actually a guy named Chris Hamilton. He is the transport driver, the man who drove the truck with Hornish's powerful Ford Mustangs secured safely inside, from Bristol to Atlanta. Remember, drivers can take two cars to every race. The second car, safely tucked away in the top of the trailer on a platform powered by hydraulics, can be used in case the first car is damaged or fails mechanically during practice or qualifying.

The #12 truck was neatly and precisely lined up and parked alongside all the other Nationwide series haulers, and Hornish's Mustang was being fine-tuned in the garage right across the walkway from the truck. In fact, in both the Nationwide and Sprint areas of the garages, the haulers were parked in perfectly straight, military-like lines, and the cars that had traveled in them were being prepared for battle in the garage bay exactly opposite the trailer. That sight was just another display of the perfect precision that seems to accompany anything that has to do with NASCAR.

On Saturday, when we found Sam Hornish's car in the garage area, we walked across to the hauler and asked for Pickle. After a minute or two, a man climbed down a ladder from the top of the truck, wiped his hands, and introduced himself. He seemed to be OK with being called either "Pickle" or "Chris," so I stuck with "Chris." He took our foursome on a personal tour of the inside of the trailer. What we saw in there was remarkable. I had seen trailers at both the Penske garage and the NASCAR Hall of Fame, but I had never seen the inside of one on a race weekend, fully loaded and ready for just about any occurrence. When we first entered the trailer, we walked into a kitchen/eating area that was so spotlessly clean it shined. Beyond that area, Pickle opened doors and drawers that housed every conceivable automobile part that the mechanics might possibly need for that particular car, on that particular track, that very weekend. Over here was a neatly organized bank of custom-built shocks, assembled just that week and constructed specifically for the track at Atlanta Motor Speedway. Over there were rows and rows of blue and red suspension springs. Farther along our walk inside, we came to the meeting room, a large room lined with black leather couches and dotted with flat-screen TVs and laptops. A conference table occupied the center of the room, and PENSKE RACING was perfectly lettered on the wall facing us. After showing us that room, Pickle directed our attention upward to an open bay. I stepped alongside him and looked up, directly into the predatory grille of the second Ford Mustang, patiently waiting its turn to come to life if duty required. A spare fuel cell rested along the other wall of the upper bay.

We thanked Chris, the mechanics, and engineers for their hospitality and set out to see what we could see before the garages became off-limits to spectators just before practice, then the Nationwide race later that day. We got a close-up look at the pink pace cars promoting breast cancer research, and we got to see both "Great Clips-Grit Chips 300" pace cars that would be used on Saturday night. We saw mechanics making measured, precise markings on treadless tires; we saw race cars up on jacks, mechanics working intently with the suspension or underbelly or engine. We saw rows upon rows of huge toolboxes inside the garage bays, something that always gets my husband's attention. The cars were being readied for battle, and every mechanic had a job to do.

At the far end of the garage area, a low steel platform had been erected. The area under the canopy was lined with long tables, and on those tables were different colored calipers and other measuring devices. Several NASCAR officials milled around the area, talking and laughing, but before long, they got down to business. There was a line forming at one end of the platform; mechanics, flanking both sides of their teams' cars, were manually rolling the cars up to the platform to be inspected. As soon as a vehicle was rolled into place, the officials went to work, using the calipers and other instruments to inspect each car before the race. One official was examining a carburetor that had been removed from under the hood of one car; another was looking underneath the car. Every team must field a car that falls within NASCAR's strictly detailed guidelines, risking anything from disqualification to heavy fines should they stray beyond those boundaries.

Shortly before practice began, we left the garage area. Only the mechanics, engineers, and drivers, plus a few spectators with special credentials, are allowed near the cars once it's time to start the engines and get out onto the track. Sweating and tired, I was ready to get back to our air-conditioned oasis for a while, anyway. On our walk back to the camper, I spotted a UPS truck with brightly colored flames licking the hood (decals, of course). It seems even UPS gets into the act when delivering last-minute packages to NASCAR teams.

On the way back, I also met and talked with the guy who runs the Huckleberry's food truck, the official barbecue of Speedway Motorsports racetracks. He kindly offered us cold bottles of water for our trek back to the camper. I could have kissed him, I was so grateful. Frances, on the other hand, stayed behind and managed to acquire about twenty driver autographs. Over-the-moon ecstatic about her achievement, she held on to that hat for the rest of the evening.

Later that night, we walked through the short roads that web the infield camping area. We saw elaborate camping setups with big-screen TVs, and campsites draped with festive strings of lights: pineapples, tires, checkered flags, and naked women. Hot and tired, we walked for about a half hour, then retired to our camping area (also festooned with strings and ropes of lights) to relax and discuss the day's events.

The very next day, I found the answer to the question I've been seeking to answer for almost a year now: "What is it about NASCAR that makes the sport such a sweeping phenomenon?" I found the reason that fans are so devoted. I found the reason that, even with changes and challenges ahead, NASCAR will not only survive, but it will thrive. I found those answers in a very unexpected place.

It's not the cars, though I will go to my grave thinking the sound of those engines roaring to life is just about the sexiest thing I've ever heard. It's not the drivers; while most of them are colorful, skilled, and interesting, they come and go. It's not the legendary parties or the scandalous bootlegging history or the raucous fans that ignite that passion that is so obvious at every race. It's love, pure and simple.

When I say I found the answer to my journalistic question about NASCAR at that Labor Day race in Atlanta, what I mean is that I had the distinct honor of meeting a gentleman by the name of Mike Fuori. Ed Clark had suggested that I meet him months ago, and although I wasn't quite sure why he made that recommendation, I trusted his judgment enough to follow through and arrange an interview. I'm glad I did. If I hadn't, I may never have sat face-to-face with the most passionate fan of not just NASCAR, but of racing itself.

A Texas meteorologist, Mike is not what I'd call your "typical" NASCAR fan. He is a soft-spoken father of three girls and husband to a beautiful wife. He is clean-cut, very well educated, and does not drink alcohol. "What?" you may ask in shocked amazement. "A NASCAR fan who is not only educated, but a teetotaler as well?" Yes, and in fact there are many well-educated fans who love racing and study the sport just as football and baseball fanatics do, and they do all that without causing themselves liver damage.

Mike was at the 2013 Atlanta race with his dad, brother, uncles (one "Drunkle" Harvey, as he introduced himself, and another named Bob). They were camping in tents in the infield, something that still impresses me. How on earth they stood the heat is beyond me. Every year, Mike and these other family members fly to Atlanta from their home states. They camp in their reserved spot in the infield—always in tents—and they immerse themselves not only in the races but also in their family history as it relates to those races.

I called Mike the week before the race, and we agreed to meet. He, his dad, his uncles, and brother came to the camper to sit down and talk with me one morning (we all decided that talking in air conditioning was preferable to talking in sweaty humidity). During the course of that discussion, I learned why Mike seems to embody both NASCAR's past and its future.

When Mike Fuori talks about NASCAR, his love of the sport is obvious; it seems almost genetic, like the color of his hair or having his father's facial features. He could talk for days—citing dates, race results, driver names and car numbers, weather conditions, and every other aspect of any particular race, but even that ability does not make him the quintessential NASCAR fan. What gives him that distinction is that every race, every story, every detail he shared with me was shared in the context of family. His grandfather went to the Daytona races in the days when they were still being run in the sand. His father, who owns a construction company, met his mother, a tool company sales rep, while in Daytona. Uncle Bob was born and raised in Daytona and hasn't missed that race since he was nine or ten years old. Uncle Harvey, who was camping at the Atlanta race for only the second time that weekend, has already started bringing his son along with him.

Bob shared a story with me about going to his first race in Atlanta, the spring race back in 1995 or 1996. He flew up from Daytona and met the other guys in his family at the Speedway. When they bedded down that night in their tents, "everything was fine," he said. During the night, they heard rain begin to patter against the tent. Then the wind picked up and the tarps they had spread over their camp site, which were held together with binder clips, began to flutter and flap. The clips started snapping off one at a time, and before they knew it, they were all standing there in their underwear trying to hold on to the tarps and protect everything that was under them. The temperature dropped like a piano off a ten-story building, and ice began forming on everything. Believe it or not, some of the nastiest winter storms ever to hit Atlanta do so in early spring. The race was postponed, as about six inches of ice had formed before it was all said and done. "That was the day I decided to become a meteorologist," Mike said.

His family's memories are as colorful as they are priceless: one year, when they found themselves without the customary fireworks they took to every race, they opted instead to make do and throw a full can of Milwaukee's Best beer into their blazing campfire. Of course, the can exploded and blew logs several feet into the air. That's the kind of story that would horrify a lot of moms. "Safety first" was always my motto as a mom, and it was probably Mike's mom's motto, as well. If I had the nerve to ask my husband and son today, or even our daughters for that matter, I'm sure I'd hear stories of similar things they had done without my knowledge. What I don't know won't hurt me, though, and therefore I will not ask. Bob told another story that had us all laughing. His dad, years ago, started the family tradition of all the guys getting together and camping at the races. "I grew up around it. It's just what we did, for as long as I can remember."

Apparently at one time it was not uncommon for fans to rent U-Haul trucks and take those to the races for camping. Back then, camping spots at the racetracks were not reserved in advance, so when the gates opened, finding a spot to camp had become a free-for-all. One year, Bob's dad rented a "bomb" of a mobile home, as he described it. "It had orange shag carpet. It was really bad," he laughed, remembering the trip. On the way to the race, the engine overheated. "We were driving down I-95, and he told one of us to keep spraying ether into the engine so it wouldn't stop running. He told us kids to stand by the doors in case we had to bail out." Again, a story that would stop a mother dead in her tracks but among guys, it becomes funny.

Bob remembers those U-Haul days as being harbingers of what he called a "big shift" in NASCAR. "People from up north were spending a lot of money to travel down here, and they were getting shut out. Then came the rule of 'no more rented trucks,' and tracks started reserving camping spots," he said, pointing to the hill that was just beside our camping spot. "See those spots up there?" he asked. "They go for about three thousand five hundred to five thousand dollars apiece." I know my mouth dropped wide open when I heard that. I could see that those spots were occupied by some very high-end, decked-out RVs, but I had no idea that the parking spots that housed them were so expensive. Yes, NASCAR has changed some over the years.

Mike recounted for me childhood memories of going to the race with his dad, playing under the grandstands, and drawing his own oval track in the dirt so he could race his small-scale cars on it. "I remember one year holding up a Ricky Rudd sign, and that year, Rudd won the race," Fuori remembered nostalgically. He told me about traveling to the race one year, only to find out about halfway there that some of the things they'd packed for camping were tumbling off the back of their truck. Their portable heater, rain gear, and even his dad's bottle of gin had rolled right off the back of that truck as the group traveled through Atlanta. "It's quite possible that we made some homeless guy's weekend," he laughed, shaking his head and smiling a genuine smile. He recounted the tale about "Hot Laps on the High Banks," a wonderful feature offered to Atlanta race fans. For a fee as low as ten dollars, fans can drive their own cars or trucks, or ride in a pace car, around the Atlanta track. They can then get their pictures taken in Victory Lane. Proceeds from Hot Laps benefit Speedway Children's Charities. "When we go to the race every year, we drive our friend Stan's old Chevy Silverado, because we build our viewing platform on it," Mike said. Describing a stroke of genius the guys had in what he called a "Hey y'all, watch this!" moment (those words always precede a reckless act by a "redneck"), they decided to drive Stan's truck, with all of their camping gear, including the heavy wooden platform, around the track. "Needless to say, the handling wasn't great." Mike then told me that, even though the truck with all that weight struggled to get around the banked turns, his dad was able to reach a speed of about 80 mph or so. But the tricky turns were too much for the weight of the truck, and his dad very nearly tipped the truck over on the first dogleg on the frontstretch. "We were all OK, and Dad pulled it out without wrecking. I'm pretty sure my brother had to change his pants when we got home, though," he said, laughing almost through tears at this point. "So what did we do? We drove around again!"

Mike then recalled for me, in remarkable detail, his favorite memory of an Atlanta race. "It was the year (Kevin) Harvick won here in 2001," he said. "Everything about that race was very special. A B-1 bomber with the number three painted on it performed the flyover. On the third lap, fans in the grandstands held up three fingers. A caution was called on lap three that day." Curious as to the significance of all the "number threes" Mike referenced, I asked him about the meaning. "Dale Earnhardt was killed in the last lap of the Daytona 500 earlier that same year. He drove the number three car."

"Oh," I said, and that was all I could muster. Earnhardt's death on the last lap of the Daytona 500 that year is still a very raw memory to many fans. I should have remembered that.

Harvick, who was competing for the Richard Childress Racing team in what was then the Busch Series (now Nationwide), was quickly picked to replace Earnhardt in the #3 car, which was in turn rebranded as the #29 car. To win the race, Harvick edged out Jeff Gordon, the driver who had emerged as being Earnhardt's fierce competition that season. Many described Harvick's win as a "healing moment" for all of NASCAR, as Earnhardt had won the race in Atlanta an impressive nine times in his career before it was so tragically and unexpectedly ended just months earlier. Harvick himself said of that 2001 win in Atlanta that it felt as though someone other than himself was making him drive better than he ever had before. He crossed the finish line in what was truly a photo finish, the nose of his car just inches ahead of Gordon's.

Fuori told that tale with such passion and, well, love, that it was difficult not to be moved by his words. Mike, who writes a blog about NASCAR, shared many of his stories with me and without exception, there were two common threads that ran through every account he gave: he remembers—in an almost uncanny manner— precise details about the races, and every single story he tells is tied to his beloved family camping tradition. "That's why it hurt so bad when they took away the second race here in Atlanta. It was just less time that we got as a family. It was one less time I got to see my dad every year."

Mike's words stayed with me for the rest of that weekend and frankly, they have ever since. Of course, he gave kudos to his lovely wife back home in Texas; not many wives would be so understanding of their family's long-standing ritual. But as I studied everything else about the Atlanta race for the rest of the weekend, my observations were cloaked in the context of Mike's perspective—love, passion, and family. That may sound silly, but it's exactly what happened. To tell you the truth, Mike's stories changed my perspective a great deal. Up until the time I met Mike and his family, I had been studying NASCAR strictly as a sport, reading everything I could and talking to drivers about the rules and the mechanics of the sport. What I had been missing all along was the "family" facet of NASCAR.

When I parted with Mike and his family after that wonderfully intimate interview, our foursome decided to pile into Steve's big Dodge pickup truck and drive through the campgrounds, just to see the sights, to see what everybody else was doing, serving, and drinking while waiting for the next race. We embarked on this journey with plenty to share with others too. Frances and I both have an affinity for sangria, so we loaded up plenty of the deep red, fruity concoction just in case anybody we met might want a taste.

Both the Nationwide and Sprint Cup races in Atlanta were night races, so that left the scorching daytime hours free for all kinds of entertainment. Shortly into our exploratory trip that afternoon, we came right up on the Fuori family campsite. Mike remembered that I had mentioned that I intended to taste some moonshine before my research came to an end. I kept hearing people talk about it; at most every race I had been to, I heard that this group had some that was "real smooth," or that group had brought some designer flavors, but I had not yet worked up the nerve to actually taste any of it. I've heard too many stories about the way moonshine used to be made; I just couldn't bring myself to do it. When we drove past their campsite, Mike waved us down and said that his camping neighbors had some moonshine to share, and they'd be happy to share some with me. As impressed as I was that he'd remembered what I had said, I was just as afraid to actually drink the stuff. In the interest of pure journalism, however, I steeled myself and worked up the nerve. Besides, how could I back out? A crowd was gathering, and the contributing host was already pouring me a shot of the innocent-looking, clear liquid fire.

Have you ever swallowed the business end of a hot poker? Downed a mouthful of tacks? If so, then you know what moonshine feels like going down. At least this moonshine felt like that; the guy poured me a shot from a vodka bottle, which gave me the uneasy feeling that maybe that moonshine had questionable origins, but of course I know that wasn't the case. I don't know the first thing about moonshine, but everyone else seemed to be indulging, so it had to be all right. Right? Anyway, I threw back the shot and felt that smidgen of clear liquid eat its way down the back of my throat, burn its way down to my stomach, and settle there to do a little more intense damage. It just sat there and blazed for a few minutes, and the suburban housewife in me swam to the surface and started panicking. I came very close to shouting to my husband in hysterics, "Call 9-1-1!" when the burning subsided and I began to think that I might be OK after all.

Of course everyone got a good laugh at my expense, as swallowing the moonshine had literally taken my breath away and made my eyes water. Frances, being the good friend that she is, took a shot of the stuff herself so I wouldn't feel alone. We thanked our host for the experience (in the South, manners are important even if a call to 9-1-1 is necessary). We hopped back into the bed of the pickup truck and continued our scenic tour, tentatively sipping our fruity sangria that we had brought along for the ride. It felt kind of daring at first, sipping sangria while sitting on the folded-out gate of a big, burly truck. After slamming back that shot of moonshine, however, the sangria seemed almost silly and a little girly.

That sunny, hot afternoon outing we enjoyed in the infield that day was an eclectic, hilarious, eye-opening experience if ever I've had one. We passed RVs—several of them—that I know cost more than our house. We passed rows of tents, some of them neat and tidy, and some of them that made tenement slums look like a step up the housing ladder. Everybody waved as we passed, and shouts of, "How y'all doin'?" came from just about every campsite we saw. The smells of home cooking and barbecuing were divine.

To our right was a pickup truck, the bed lined with a blue plastic tarp and filled with water. It was a mobile swimming pool, of course, and there were two very pretty girls sitting in it and drinking beer. Up ahead were a couple of retired school buses-turned-campers, with NASCAR beach towels covering the windows—either the owners' attempt at shading the interior or a feeble shot at ensuring some privacy. Cornhole games were set up everywhere, and people laughed, talked, and drank while they played the game. One guy was actually stepping off the distance between the two boards, trailing his measuring tape behind him. That was going to be a serious game of cornhole.

Turning a corner and heading down another short stretch of road in the infield, we came upon a group of young boys playing a game. They had set up a long table that had several Solo cups arranged on it, and they were bouncing a Ping-Pong ball on the table in an attempt to get the ball to land in one of the cups. Yes, I have just described for you the game of "beer pong," and the boys who were playing looked to be no more than ten years old. I did not stop to ask them what was in the cups; I figured if their parents were there, that was their business. At one campsite, we came across two couples who were, I'd say, in their mid- to late-sixties. One of the women was wearing a pink and black zebra print miniskirt and black tank top, and the other was decked out from head to toe in gold lame and layer upon layer of costume jewelry. Both women were tanned almost beyond recognition. Their husbands, who were busily cooking up some delicacy or another on their portable grill, were both shirtless, wearing plaid shorts, black dress socks, and sandals.

A few rows over, we came across an old revamped school bus with a wedding arbor and chairs arranged on a platform that had been mounted on the top of the bus. Checkered flag garland, draped around the entire setup, fluttered in the occasional tired breeze.

A candelabra held five white candles like so many long skinny fingers, and I had the strangest thought: *I wonder how long it'll be before those candles melt into a greasy, waxy mess?* A couple from Terre Haute, Indiana, got married right there on top of that school bus that day, and Brother Bill himself had performed the ceremony in front of about 150 onlookers. The guests weren't there by way of formal invitation, mind you. They just happened to be in the area, and they wanted to toast the couple and wish them well.

By far the most outlandish thing we saw on our afternoon tour was a young woman, about eighteen years old, sitting in the bed of an old pickup truck with what appeared to be her grandparents, just passing the time until the race would start later that evening. The young woman was wearing a bright green tank top, and she had one side of the top pulled down to expose an imitation rubber breast. She was drinking beer from a can that she held in one of those giant fake monster hands one might see in costume shops before Halloween. I still can't figure out what the purpose of that exhibition was. I can't figure out why she'd do such a thing with her grandparents sitting nearby. She was getting quite a lot of attention, so I suppose that was the reason. I don't think I'll ever get that image out of my head, though. What a bizarre thing to do; I think I would have been less shocked to see her flaunting the real thing peeking out from underneath her tank top than that grotesque-looking rubber imitation. I'm still kicking myself for not taking a picture of the whole display, but I think I was, quite literally, in shock.

Most campsites we passed housed good-natured campers that ranged in age from very young children to white-haired grandpas. They would be sitting at a table playing cards and eating, or they'd just be relaxing in lawn chairs watching the goings-on around them, passing time until the green flag was waved to start the race that night. But note what I said about the people we saw that afternoon. I think I picked up on it because of my conversation with the Fuori family; what I saw were families, generations, all together for an entire weekend. Hanging out. Talking. Laughing with one another. Yes, some of them appeared to be weird and outlandish, and some redefined the word "redneck" as I previously understood it, but they were all happy to be there. I'll tell you something else too: not a single one of them gave a flip about what I thought. They were there to have a good time. I didn't see any teenagers rolling their eyes or off sulking darkly by themselves. I didn't hear anyone griping or complaining, and I saw no one fighting. These were families, generations of them, all living together, however briefly, to enjoy a shared passion. Young ones were learning that passion from their elders, and the elders were more than happy to talk to anyone who wanted to listen about the "old days" of NASCAR. Where else can you find three, sometimes even four, generations of family all gathered with the same interests?

I saw food everywhere: casseroles, dips, cakes, cookies, and pies, and we didn't mingle with one single group who didn't offer us some of their homemade bounty, and we in turn did the same. It's just the way things work in NASCAR tailgating.

Our campsite was in a secured area of the infield, and there was a guard posted at the entrance all day and all night. His name was Shannon, and he was the nicest guy you'd ever hope to meet. We chatted with him every day as we headed out to explore, and we chatted with him every evening when we returned to get ready to watch a race. When he had a break, he'd come to our campsite and just hang out with us. We took him dinner every night, and he was grateful to get it. Of course, he could have eaten at any spot in the entire track he chose while on duty, but there was just something special about our bringing him a heaping plate of whatever we had cooked that evening. The morning after we had treated him to a steaming helping of spicy low country boil, word of the delicious meal had spread throughout the vast network of security guards. Everyone wanted some of that fiery South Carolina delicacy—or hamburgers, or Mexican food, or whatever we had prepared that evening.

It occurred to me during that unforgettable Labor Day weekend in Hampton, Georgia, that NASCAR seems to be the great leveler. There were corporate executives, construction workers, teachers, homemakers, even a meteorologist camping side by side, sharing food and drink, and all there for the same reasons—cars, drivers, friends, and family—and a great big party.

On Saturday night, we prepared to watch the Nationwide race. We had seen the practice runs, and we had watched qualifying runs, but we were all ready to see a race. We had grilled hamburgers for dinner, taken Shannon his plate, and cleared the dishes. It was time to take our seats. At this race, our seats were on top of our camper, and there were no better seats in the house. By dusk, the temperature had mercifully dropped below ninety degrees, and there was actually a welcome, faint breeze. The night sky was beautiful.

Steve had brought along a ladder for the climb up to the roof; although the camper came equipped with a narrow ladder mounted almost flush with the outer wall, both of our husbands thought it wise to bring along a more accommodating one. This one leaned at a more comfortable angle against the camper, making it easier for a woman still recuperating from a couple of knee surgeries to navigate. We all climbed up to the roof of the camper and found a comfortable spot with a great view of the track and all its turns. The guys had already hauled a cooler up there, loaded with beer, soft drinks, sangria, and water. I had to admit that, despite the climb that I had been dreading for days, this was the best seat I had ever had at any race.

Kevin Harvick won the Great Clips-Grit Chips three-hundred-mile race that Saturday night, with Kyle Busch and Sam Hornish taking second and third places, respectively, in the contest. When the race was over, we wearily climbed down the ladder, attempted to wash the day's sweat, grime, and flecks of rubber off ourselves, and collapsed into bed. I slept better that night than I had in weeks in my own comfy bed at home. Exhaustion really is a wonderful sleep aid.

We spent the better part of Sunday doing the same things we had done on Saturday. We toured the garages again, and on Sunday, the level of activity was almost feverish. There were a lot more people milling around in that area too. Later, we were all thrilled when Ed Clark took time out of his hectic race weekend schedule to stop by our camper and say hi, making sure we had everything

we needed. While I stayed behind to decipher some of the notes I
had hastily scribbled during our Saturday adventure, my husband
and friends spent time outside watching the ever-increasing buzz and
activity in the infield. By Sunday afternoon, the middle of Atlanta
Motor Speedway was completely packed, elbow-to-elbow with
campers. The level of activity inside the track had increased as well.
Golf carts zipped around, carrying very important-looking people
to their very important destinations. Infield campers milled about,
trying several different locations to see which afforded the best
view of the race. At precisely 7:30 p.m. that evening, forty-three
of the most talented drivers in the world would be competing in the AdvoCare 500 to win NASCAR's Sprint Cup race at
Atlanta Motor Speedway, and no one wanted to miss the spectacle,
the competition, the excitement of what was about to happen.
The stands were fast filling up with fans.

About an hour before the race was scheduled to start, a light
rain began to fall. It looked as though a delay was a sure thing,
but those handy track dryers were brought out to make their slow,
deliberate rounds on the track, thereby allowing the race to start on
time. Professional photographers were setting up their equipment;
an ESPN photographer had established his territory just fifty feet
from our camper. The pyrotechnics crew had also set up shop very
near our site, and of course both Steve and my husband spent time
talking with them. How could any man resist having a conversation
with guys who blow things up for a living? The crew let them in
on the planned timing of the first fireworks display, an impressive
succession of red, white, and blue rockets timed to coincide with
the first wide-open lap of the Sprint Cup race. It was an impressive
sight, and we caught it all on camera, thanks to a friendly insider tip.

Two guys from the county water department were monitoring
the water pressure at a hydrant just outside our front door, and
by race time, even they had climbed up to the roof of the camper
to watch the race with us. Shannon took a break from his post to
check out the view from our perch, and of course at our invitation,
he helped himself to a plate of food and took it back to his security
post, where he enjoyed the meal during the prerace festivities.

The University of Georgia's chaplain Kevin Hynes delivered the invocation that evening, and Ernie Haase and Signature Sound sang the National Anthem, which was followed by a Cessna Citation flyover. AdvoCare's President and CEO (and husband and wife team) Richard and Sherry Wright delivered the command to racers to start their engines.

This would probably be a good time to inject another observation I've made about going to see NASCAR races live. You'd better say everything you have to say before the call to start engines, because for the next five hundred miles (325 laps at AMS), all conversation stops. Texting takes over as the preferred mode of communication during a race; if you don't text or use sign language, whatever's on your mind will just have to wait.

Frances taught me how to keep up with the minute details of a race that you just can't catch while watching it live. During the races, she looks at her phone almost the entire time, reading her Twitter feed. In fact, about half the fans in the stands were looking from the track, to their smart phones and back once the race started. Fans, driver teams, and racetrack staff tweet constantly during races. In fact, I read about "the slap heard around the world" on Twitter.

NASCAR's Camping World Truck series race was held not in Atlanta on that weekend, but for the first time ever, in Canada on Sunday. The 3.957-kilometer Canadian Tire Motorsport Park road course was, according to some, the perfect venue for the Camping World Truck series race. Here's another first: Driver Mike Skeen's girlfriend lost her cool after that Sunday race and slapped driver Max Papis right across the face. Since she had signed in to the race as part of Skeen's team, NASCAR fined her $2,500 on the following Wednesday and banned her indefinitely from all NASCAR events. Skeen's crew chief was also fined $2,500 after the slap, because it's his job to keep all team members in line. Why the skirmish? During the race, Papis and Skeen were battling for third place during the last lap, then the cool-down lap, of the race. Apparently the girlfriend didn't like what she saw when the two drivers "swapped a little paint" as they jockeyed for better finishes.

I read all of that on Twitter, by the way, during the Sprint Cup race.

During that Sunday night race in Atlanta, Bowyer's Toyota engine blew up after about 190 laps, squashing any hopes he had for a win there. He had already clinched a spot in the Cup Chase, but it was still a frustrating turn of events. Kyle Busch ended up winning the race on Sunday night, securing himself a spot in the Chase as well. When he crossed the finish line first, my Twitter feed started rolling so fast that I could barely keep up. The Busch brothers are a colorful pair in NASCAR; fans either love them or hate them, and rarely is there any in-between.

I must say, that weekend at Atlanta Motor Speedway in Henry County, Georgia (not far from where Scarlett O'Hara fluttered her eyelashes at every beau she met in *Gone with the Wind*), taught me more about NASCAR than I could ever have imagined. I got to see for myself what makes the fans tick. Why would thousands of people brave Georgia's August heat for days, just to watch a race or two that will be televised? There were at least twice as many people camped outside the racetrack as there were inside, mind you. They were there for the races, yes, but they were there because of family too. They were there because of the legendary party. They were there because Dad was a Petty fan, or an Earnhardt fan, or just a longtime fan. They were there to watch competitors who still seem like the guy next door, like "regular guys," and they imagine, if even for a split second that they, too, are incredibly skilled drivers who happen to be competing for a nearly 6.5 million dollar purse, beautiful women hanging on their every word and fans clamoring for their autograph. NASCAR is the last sport out there in which that could still realistically come true for a lot of people.

I thought that race in Atlanta, which nearly brought to a close the regular season of 2013, would be the last big story I'd chase in this journey.

And then came the Richmond race.

Chapter Nineteen
Define "Fair and Square"

The last Sprint Cup race of the 2013 regular NASCAR season was held on Saturday night, September 7, at Richmond International Raceway in Virginia. Many sports analysts intimated that the race was inconsequential to all but a few drivers who were close, but had not yet secured a spot in the Chase for the Cup. Boy, were they wrong.

That evening's race ritual had started just as every other final race of the season had, with a prerace meeting in which drivers were reportedly reminded by NASCAR officials to "play fair and square." For many years, as far as I know, that's precisely what they've done. But during the final few laps of the last regular season race of the year, events took place on the track that may very well have changed the rules of NASCAR forever. To NASCAR purists and devoted race fans everywhere, the Richmond race shenanigans and domino-effect results were nothing short of mayhem there for a few days. Even the top NASCAR brass had to take a few days to sort things out before handing down decisions and changes to the official NASCAR rule book.

With seven laps to go in the four-hundred-mile race that Saturday, Clint Bowyer spun out for no apparent reason in his #15 car, thereby instigating a caution that interrupted the momentum of all the drivers but specifically, that of driver Ryan Newman. Newman, who was leading when the now-famous spinout happened, dropped to fifth place after the caution. Driver Brian Vickers, one of Bowyer's teammates, also appeared to inexplicably slow down during the final laps of the race. That action allowed a third Michael Waltrip Racing teammate, Martin Truex Jr., to beat Hendrick Motorsports' driver Jeff Gordon by one point in regular season standings. It's convoluted and complicated, but this succession of actions and suspected collusion of Michael Waltrip Racing and two other teams essentially cost Newman and Gordon a shot at a berth in the Chase. At first.

Fans and commentators alike were abuzz over the seemingly blatant manipulation of the Richmond race outcome, and a review of the teams' radio communications did indeed raise some serious questions about manipulation of the Richmond race results. NASCAR sat up and took keen notice, and on the Monday following the race, the sanctioning body took unprecedented action in response. NASCAR handed out fines and penalties as though they were high school principals handing out after-school detention slips.

Truex was stripped of his playoff spot, and Newman was granted a spot in the Chase. Bowyer and his other driver teammates were fined 50 racing points (along with some financial penalties), and team owner Michael Waltrip was fined a blistering three hundred thousand dollars. Penske Racing and Front Row Motorsports were placed on probation for the remainder of the 2013 season, though those teams' drivers, Logano and Gilliland, were not accused of any specific wrongdoing. As I said, the matter is complicated and difficult to follow, even for seasoned race fans. There's never been another series of incidents and penalties like this one in the history of the sport.

The cherry on top of the cake, after further consideration by NASCAR, was the unheard-of addition of a thirteenth driver to the Cup Chase: Jeff Gordon. Again, a history-making move.

Say it isn't so, Clint. Did the driver I chose way back at the beginning of the season cheat at Richmond, thereby helping his teammate advance to the Chase? It's no secret that he and Gordon did not end the 2012 season as best friends, but cheating? I've been wearing T-shirts with his name plastered all over them for months now. *Clint, did you cheat?*

That all depends on whom you ask. There's a very fine line in NASCAR between racing to benefit your team and playing "fair and square." So what the heck happened in Richmond? Radio communications, that's what. Remember that a driver talks with his spotter in the stands and other teammates constantly throughout a race. He tells his team if the car is not handling properly. He reports any strange or unusual noises or vibrations. He hears from his team-mates as to who's where on the track, and what's going on with other drivers and their cars. Apparently, those same radio commu-nications can be used to wheel and deal with other drivers during a race, and that's what NASCAR suspects happened in Richmond.

Now it's not exactly news that drivers will get help or give help on a track, and that help is sometimes brokered, so to speak, over those radios. Is that cheating? In the purest sense, yes, if drivers are expected to race at their full-out best from green flag to checkered flag. But why be on a team if the team can't act as, well, a team? If a slow lap here, a late-race pit there, and a victimless spinout over here can bump a driver up in standings, isn't that just savvy team-work, pure and simple? Maybe. The problem, it seems, with the actions that took place in Richmond is that all those who were involved seemed to be pretty casual and careless about what was said on the radio. In other words, some drivers were suspected of working together to manipulate the outcome of the race, and that had fans questioning both the integrity of the sport and the honesty of some drivers. NASCAR, ever vigilant about the integrity of the sport, was not happy about that at all.

In a September 14 media conference, NASCAR Chairman and CEO Brian France, along with NASCAR President Mike Helton, laid out the new rules after they had been presented to drivers, owners, and crew chiefs in a closed-door meeting earlier that day. According to a statement made by France in the Saturday media conference, the new rules "are all designed to do what our fans expect, and that means that their driver and their team give one hundred percent to finish as high up in a given race as possible."

Helton read the new rule verbatim as it will be listed in the official rule book but basically, drivers are expected to race to win, and any attempts to alter or manipulate the outcome of a race will be punishable by fines, points penalties, probation, suspension, or a combination of all the above. Both the "manipulating" driver and the "beneficiary" driver are subject to those consequences, just to cover all the bases.

Now that the water's even murkier, I must say that once again, NASCAR responded swiftly to the Richmond shenanigans, and I believe they made it very clear in that news conference that they expect drivers to full-out race during every competition and not play games. But even now, the question of unfair moves on the track will likely remain a subjective one, just as there are unclear rules and motives in other sports.

For instance, when a quarterback sees several three-hundred-plus-pound defensive linemen descending on him like ducks on a June bug, is it cheating to throw the football to an area where there is no receiver, simply to get the heat off himself, thereby avoiding a sack and a lot of pain? If you're a football purist, sure it is. It may not be called outright cheating, but the quarterback is no doubt trying to manipulate the outcome of that one play; specifically, he doesn't want to get pounded into the ground by guys twice his size and lose yardage or get sacked doing it. In my opinion, that's just smart play, for both himself and his team. If the officials don't like what they see though, they call it "intentional grounding," but to me, getting rid of the ball is just plain old common sense. You see, it's all a matter of perspective.

NASCAR fans, as always, offered their opinions freely and honestly in reaction to Bowyer's spinout and everything else that allegedly took place in Richmond. Their responses ranged from full agreement with NASCAR, to "they didn't do enough to stop it," to "Aw, let 'em race." Fan Dennis Lynn thought that both Bowyer and Logano should have been docked enough points to knock them out of the Chase, even though he is a staunch Michael Waltrip Racing supporter. Fans Susan and Rick Chappalear echoed Lynn's sentiments, adding that knocking those two out of the Chase would have made room for Newman and Gordon, thus eliminating the need to have a thirteenth driver in the 2013 Chase. All of these fans felt that Michael Waltrip Racing should have been fined five hundred thousand dollars, not three hundred thousand dollars, to atone for the grievance leveled against the organization. Frances, my friend who camped with us in Atlanta and coached me through this entire season, weighed in on the matter, as well. "The fact that NASCAR stepped in and penalized all MWR drivers was a bold move. It was also a move that needed to happen. Multi-car teams can outmaneuver single-car teams, so NASCAR needs to police this going forward." Perhaps the most succinct opinion about the whole ruckus came from fan John Vaughan: "Earnhardt would have done it." To a fellow NASCAR fan, that one sentence speaks volumes.

Other fans thought that all the hullaballoo was "much ado about nothing," as my longtime friend Bill Shakespeare might have said. "Let the guys race," said Don "Hootie" Hershey, a racing fan from Louisiana that I met in Alabama while researching this historic turn of events. "So what, if the guy spins out in a race, it's his own neck. He didn't hurt anybody." Laughing, Hootie added, "There ain't no crying in racin'."

Chapter Twenty

The Phenomenon of NASCAR, As I See It

It has taken me nearly a year to familiarize myself with something I've been hearing about all my life, and that's NASCAR. Why did I do it? Well, because I truly did want to learn about it firsthand. Too, I really wanted to understand how something that seemed to be so boring, so one-dimensional, as I believed racing to be, could attract the people and the money that it consistently does. I may have formed an opinion (however uninformed) about NASCAR over the years, and yet my curiosity had been piqued by the wild popularity of the sport.

How will this remarkable journey impact my pursuit of knowledge about the world of NASCAR? I know without a doubt I'll go to more races, and I'll always stay tuned to the latest buzz as the sport evolves, and evolve it will, make no mistake. NASCAR is not run by a klatch of ignorant hillbillies, and it doesn't generate the money that it does by being stagnant and unresponsive to fans' desires. It will change as long as it exists. Safety, technology, integrity, and yes, even the fans, will drive that change.

I will say this: I now understand why people love it. I can see how fans get wrapped up not just in the races, but also in the goings-on of the drivers, the teams, and even the most powerful men at the top of the organization, and they aren't afraid to tell you exactly how they feel about any of them. I understand their passion, and not to belabor the point, but I certainly do understand the pull and attraction of the cars themselves. Have mercy, I do love the way those cars sound, especially when those engines turn over for the first time. It probably wouldn't do for me to ever be the person who commands the drivers to start their engines. I'd almost surely faint dead away when they did, microphone still in hand.

NASCAR and its fans have been stereotyped for many years. From the old-school sponsors (beer and cigarette companies headlined the races for years) to the fans who, admittedly, sometimes go a little crazy at races, that stereotypical reputation for being "rednecks" has been hard-earned. Never forget that, to a redneck, being called a "redneck" is nothing short of a compliment. As with everything else that is NASCAR, however, I think that fan reputation will evolve. I think the NASCAR fans of the future will still love a good time, but I also think they will begin to look a lot more like the Mike Fuoris of this world, and less like the stereotype that many outsiders picture when they hear "NASCAR."

Will we still hear shouts of "Hey, show us yer boobs!" emanating from the grandstands and camping areas? Probably. I said the fans would change, not die off as a species. As long as there are women willing to oblige that call, there will be men willing to issue it. That's not really a NASCAR issue, it's a man/woman/respect/self-respect issue, but that discussion is not for our purposes here.

Will we still see fans at races getting so blindly drunk that they likely won't remember the race for which they bought tickets? Of course we will. Have you ever camped for three or four days or more, especially in a tent? Drinking gets to be one of the only pastimes that's entertaining after a while. In fact, I'd worry more about someone who camped for that long and didn't eventually start drinking, even a concoction as girly and tame as sangria (unless Frances makes it). I think some things about the fan base of NASCAR will never change and that's OK, because they are perfectly fine just like they are. I think too, however, that some characteristics of future fans will most definitely transform. As Mike Fuori put it, and he speaks passionately about the future of NASCAR, "Fans want to be comfortable. They want to know when they get to a race, even if they're camping, that they're going to have a Wi-Fi connection. They want to know that there will be fun things for their kids to do at the races. They want to know that they can take their family and not worry." In other words, the fan of the future will be a tech-savvy family guy. That's not so bad, is it? We're raising an entire generation of tech-savvy people right now. And those tech-savvy consumers are going to have jobs, and money, and passions. Why shouldn't NASCAR be one of them?

Gone are the days of the great warriors of the track—Junior Johnson, Rex White, Richard Petty, Dale Earnhardt, Bill Elliott, and so many more—the drivers that fans supported no matter what, all the time. Those drivers and those who are in that same category with them really were considered to be larger-than-life legends, and throngs of adoring fans followed their every move on the track, and supported them, win or lose. It's probably even safe to say that the days of fans supporting a particular make of car—Chevys, Fords, and what? Toyotas?—are long gone. Yes, the NASCAR landscape is changing, and it appears to now include just the tip of Japan. For goodness' sake, some fans still can't get past the thought of a Toyota racing on a NASCAR track, so they surely can't get behind the drivers who race them.

But dawning are the days that will see fans who are fans strictly because of the awesome mechanical and engineering technology that is spawned and raised in NASCAR. You'll find fans of teams and strategic alliances at races now. Fans of more than one driver are common nowadays too—fans who admire the combined characteristics of two, three, or even more drivers. There are fans of actual races, of the events themselves. There are fans of certain tracks; Talladega is the first that comes to mind. As the man said at the beginning of this year's May race, "This is Talladega," and fans love what that means, historically. It means wide-open racing on a superspeedway, even with restrictor plates safely in place to govern top speeds. It means an all-out raucous, sometimes even bawdy, party. It means spinouts and wrecks that are likely to be spectacular in scope; the long tracks bring out the best of the worst of NASCAR smashups and crashes, and fans love it. I'm still working on that part; I haven't gotten to the point that I can cheer for a nasty wreck, even if no one got hurt. But I'm still a little new at this, so give it time.

I also believe that, as much as NASCAR fans of tomorrow will likely evolve into some very technologically capable, well-educated aficionados, there will be some things that will never, ever change. One of those stronghold characteristics is patriotism, exhibited so unabashedly at races. NASCAR grants high honor and praise to our nation's military, and so do its fans. Fans will never stop standing and placing their hands over their hearts when the National Anthem is played, and they will always understand what those demonstrations of respect mean. They will always look overhead during the last few notes of that song for the flyover, federal belt tightening be damned. They will always feel chills when they see it.

NASCAR fans will always gladly bow their heads before every race and pray along with that race's presiding pastor or chaplain or just plain old preacher, and that prayer will always be a Christian prayer to the Christian God. No one on the track or in the stands will be ashamed of that fact. They will always cheer after that prayer has been offered up.

Yes, NASCAR will change. Many fans don't like to hear that, but they should take heart in the fact that in order to survive, NASCAR has to change. Its leaders believe that. That's why the rules change, because the drivers' abilities are changing, improving. That's why there's a "hashtag-NASCAR" on Twitter. That's why the tracks and drivers and teams have hashtags and handles and Facebook pages and who knows what else. Will it change because the sport's longtime fans don't matter any longer, or because NASCAR has forgotten where its roots lie? Absolutely not, at least not in my opinion. NASCAR will change to survive. It will change so that younger fans will view and participate in a manner in which they're comfortable, more at-home. And those newer, younger fans will develop passion and love for NASCAR and its players for their own reasons. Of course, I believe they'll still go to the races, because there is absolutely nothing in the world that compares to the sights, smells, and that glorious sound of forty-three powerfully engineered, gasoline-fueled works of art lunging to life to do battle. No technology, not television or smartphones or tablets, can come close to what that sounds like in real life, and I love it.

I searched and searched for some closing words that would help me communicate what NASCAR's all about and why I believe it has swept through this country and many others with a fury and a passion like no other sport I've seen. I wracked my brain; I went back through all my notes and interviews, trying to find that one quote, that one line that would be both clever and characteristic of NASCAR. What I came across was not just one line or a pithy quote; rather, it was a sobering and heartfelt prayer delivered by Pastor Joe Nelms of Family Baptist Church in Lebanon, Tennessee, right before the 2011 Nationwide Federated Auto Parts 300 NASCAR race in Nashville, Tennessee. With that prayer, the pastor drew both praise and criticism, with some here in the Bible Belt going so far as to call his spirited invocation "blasphemous." Nelms said he was simply trying to deliver a prayer much like the Apostle Paul might have back in New Testament days, and he gave it his best shot. I can't say what Paul might have thought of it—I hear he was quite a talker—but I think Nelms hit his target dead center when he prayed to the Lord above:

"Heavenly Father, we thank you tonight for all your blessings. You said in all things give thanks, so we want to thank you tonight for these mighty machines that you've brought before us. Thank you for the Dodges and the Toyotas. Thank you for the Fords, and most of all we thank you for Roush and Yates partnering to give us the power we see before us tonight. Thank you for GM Performance Technology and the RO7 engines. Thank you for Sunoco Racing fuel and Goodyear tires that bring performance and power to the track. Lord, I want to thank you for my smokin' hot wife tonight, Lisa. And my two children, Eli and Emma, or as we like to call them, the 'little E's.' Lord, I pray you bless the drivers and use them tonight. May they put on a performance worthy of this great track, in Jesus' name."

And how did Nelms end this priceless petition to the Good Lord Almighty on behalf of the drivers, teams and fans? Why, the only way he could, of course.

"Boogity, boogity, boogity, Amen."